Francesco Cester

Manfred Clemente

Hans-Michael Korff

Newton and Relativity

—

An alternative approach to relativistic mechanics

by means of the Lex Secunda

Mutationem motus proportionalem esse vi motrici impressae,

et fieri secundum lineam rectam qua vis illa imprimitur

Francesco Cester

Manfred Clemente

Hans-Michael Korff

Newton and Relativity

—

An alternative approach to relativistic mechanics

by means of the Lex Secunda

Bibliografische Information der Deutschen Nationalbibliothek:

Die Deutsche Nationalbibliothek verzeichnet diese Publikation

in der Deutschen Nationalbibliografie; detaillierte bibliografische

Daten sind im Internet über http://dnb.dnb.de abrufbar.

Verlag:

BoD · Books on Demand GmbH, Überseering 33,

22297 Hamburg, bod@bod.de

Druck:

Libri Plureos GmbH, Friedensallee 273,

22763 Hamburg

ISBN: 978-3-8192-3052-3

Preface

There are two types of publications on the Theory of Special Relativity.

To the first category belong technical works, which make use of mathematics to rigorously prove relativistic laws.

These works are reserved for readers who have the mathematical and physical knowledge necessary to understand their scientific content.

Publications in the second category are intended for those who are interested in the theory but lack the necessary notions to grasp its mathematical formalism.

The authors of these popular works promise to explain the concepts in a way that is understandable even to non-experts in the subject.

All works, whether of the first or second category, have in common the partly paradoxical assumptions on which the theory of relativity is based.

Readers who have difficulty accepting these assumptions very often remain skeptical of the theory.

The present work does not belong to any of these categories.

The present work initiates the relativistic discussion on the basis of clear and logical assumptions; with the aim of overcoming what the author believes is an overly restrictive interpretation of the field of validity of Newtonian mechanics.

In fact, since the development of the Theory of Special Relativity more than a hundred years ago, there has been consensus among scientists worldwide that Newton's physical laws have only limited validity.

The commonly accepted restriction is that Newtonian mechanics based on these laws is only applicable to constant masses and low velocities.

However, the second law of motion, correctly defined in the sense of Newton as a time derivative of the momentum, consists of two terms. The first well known term describes the speed change. The second generally unused term takes into account a possible change in inertia of the body on which the force acts.

This work shows that by using both terms, Newton's second law of motion remains valid at variable mass and high speeds, thus setting the stage for a new interpretation of Theory of Relativity.

The demonstrations given here bring the physical arguments back to a simpler and more intuitive level than that of the traditional relativistic interpretation, thus have the merit of making them understandable and obvious.

The work concludes with the section "Historia operis" in which the author explains the methods used to calculate the derivations.

Elementary knowledge of mechanics, particle physics and of the Theory of Special Relativity is sufficient to understand the mathematical requirements.

To conclude this preface, I would like to point out that the contents of this book are also available from the same publisher in Italian under the title "Newton e la Relatività" and in German under the title "Newton und die Relativität".

Contributors

This work is the result of the collaboration of three persons:

Francesco Cester carried out the calculations of the derivations and dealt with the physical contexts. Manfred Clemente and Hans-Michael Korff have checked the mathematical and physical consistency of the work and have edited and translated the text from German.

List of Contents

Introduction

In the second half of the 19th century, the Scottish physicist James Clerk Maxwell unified electrical and magnetic phenomena in the theory of electromagnetism, describing its fundamental laws with a system of four differential equations.

This was a hugely important achievement for physics.

However, while this achievement made physicists aware that, with Newton's laws for mechanics and Maxwell's equations for electrodynamics, science was finally able to explain all natural phenomena, it also left them in a state of uncertainty.

In fact, Maxwell's equations proved to be incompatible with Galileo's transformation on which classical kinematics is based.

The transformation suitable for Maxwell's electromagnetism was calculated by the Dutch physicist Hendrik Antoon Lorentz. This transformation involved not only the spatial coordinate but also the temporal one.

Solved for empty space, the Maxwell's relations reduced to the equation of a wave propagating at the speed of light. This fact supported the hypothesis of the electromagnetic nature of light.

With Michelson's interferometer in 1886, the constancy of the speed of light was demonstrated for all inertial frames of reference. This experiment confirmed the insufficiency of classical kinematics and at the same time manifested the need to develop a new physical theory.

In 1905 Einstein, based on the constancy of the speed of light, derived the transformations that correctly describe the behavior of space-time at high speeds. These transformations are the same that Lorentz had calculated, adapting them to Maxwell's equations.

Lorentz's transformations, replacing Galileo's, eliminated the incompatibility of electromagnetism with kinematics and laid the foundation for a new theory: the Theory of Relativity.

The immediate consequences are:

- Relativity of simultaneity: Simultaneous events for one observer may not be simultaneous for another in relative motion.
- Time dilation: Moving clocks appear slowed down compared to "stationary" clocks.
- Contraction of lengths: Objects appear to shorten in the direction of their motion.
- Speed limit: No physical object can exceed the speed of light in a vacuum.

Space and time, therefore, can no longer be considered absolute, as Newton believed.

These are the assumptions on which the theory is based.

Those who are reluctant to accept these assumptions tend, more or less unconsciously, to take a skeptical attitude toward the theory itself.

I, too, have been in this situation for a long time, so I have sought another approach to the theory of relativity that starts from intuitively accessible premises.

This intent was later realized in the following paper, the stimulus for whose writing was given to me by a scientific article that appeared under the title "An elementary derivation of $E=mc^2$"[1] (see Appendix A I).

---

[1] Published 1990 in the American Journal of Physics, on page 348, Volume 58, Issue 4.

With this article the physicist Fritz Rohrlich was able to show that Einstein's famous equation $E = mc^2$, which describes the equivalence of mass and energy, can be derived from the laws of classical physics without using relativistic mechanics.

The proof, based on the Doppler Effect for electromagnetic waves, shows that Einstein's famous formula does not necessarily presuppose the Theory of Relativity.

The question that now arises is this:

Under the assumption that the equivalence of mass and energy can be derived from classical physics, is it then possible, based on this principle, to introduce a more accessible approach to the theory of relativity than that which has been established since the beginning of the 20th century?

The established method of derivation, based on the postulate of the constancy of the speed of light for all inertial frames of reference and on the Lorentz transformations, rejects the concept of an absolute space and an absolute time and is therefore not easy to grasp.

Is it then possible to follow another, easily comprehensible path to prove the relativistic laws?

Through my research I have found that, starting from Newton's second law of motion in conjunction with the principle of equivalence of mass and energy, it is possible to pursue a new approach that establishes a direct connection between classical and relativistic mechanics.

The present study shows the results of these investigations. The work is divided into three parts:

Description of the New Methodology

The first part examines the scope of the second law of motion (Chapters 1 and 2) and presents three alternative proofs of the equivalence principle of mass and energy (Chapters 3 and 4).

In the second part, Newton's second law of motion is used in conjunction with the Mass-Energy Equivalence Principle to derive the dependence of mass on velocity (Chapter 5) and to extend the calculation of kinetic energy to high velocities close to the speed of light (Chapter 7).

These two principles then form the basis for further derivations during the third part of the thesis.

Within the framework of thought experiments and by applying the law of conservation of energy and momentum, the relativistic addition formula of velocities can thus be derived without using the Lorentz transformations (Chapters 9 and 11).

Chapter 14 concerns the central objective of the work.

Among the phenomena of nature currently considered incompatible with Newtonian mechanics in the physics world, the constancy of the velocity of light in vacuum, which can be demonstrated for all inertial frames of reference, is the most important one.

The constancy of the speed of light is the central postulate of the theory of relativity and stands symptomatically for the gap between classical and relativistic mechanics.

Chapter 14 nevertheless shows that the theoretical confirmation of the constancy of the speed of light can be achieved with the help of Newtonian mechanics.

Finally, in the last two Chapters, the kinetic dependencies of the frequency of the electromagnetic radiation (Chapter 16) and of the acceleration (Chapter 17) are derived.

For the derivation of the formulas, only the conservation laws of energy and momentum are used.

Results of the New Methodology

As a result of the entire study, the following conclusions can be drawn:

- The equivalence energy-mass is not necessarily a relativistic principle but, as Einstein himself proved (see Chapter 3), a principle derivable from the laws of classical physics.

- The constancy of the speed of light in vacuum, which is valid for all inertial frames of reference, is not a postulate to be assumed a priori, but a principle provable by the laws of physics.

- Using the second law of motion as its fundamental principle, the theory of relativity can be considered a logical extension of Newtonian mechanics.

Finally, Newton's laws of dynamics form a broader physical basis than is usually assumed.

1 About the Limitations of Classical Mechanics

The limitations of classical mechanics in explaining natural phenomena are mentioned at the beginning of many works on the theory of relativity.

Similarly, these works often proceed as follows.

Most treatises are based on experiments that prove the constancy of the speed of light in all inertial frames of reference, regardless of their state of rest or motion.

Since observers in different frames of reference measure the same speed of light despite their relative movement to each other, they cannot agree on the simultaneity of events or the dimensions of bodies.

Based on these findings, one is forced to abandon the idea of an absolute space and an absolute time.

However, since space and time are relative quantities, there arises the necessity of redefining the relativity principle of mechanics.

As a result of this development, coordinate transformations between frames of reference[2] and the measurements of space and time observed within them, are determined as a function of their velocities.

These hypotheses are then confirmed to be consistent with experimental observations at high speeds.

Finally, attention is drawn to the limitations of classical mechanics in both the cosmic domain and the experimental field of subatomic particles.

---

[2] These are the so-called Lorentz transformations. They fulfill the central requirement of the Theory of Relativity: the same physical laws must be valid independent of the relative speed between the frames of reference.

This is essentially what is said in the publications on the theory of relativity about classical mechanics.

The criticism is mainly directed at Galilei's transformation and thus reveals the inapplicability of classical kinematics to high speeds.

In order to fully analyze the limitations of classical mechanics, however, it is necessary to determine whether this insufficiency of kinematics is accompanied by a similar deficiency of classical dynamics.

Let us see what conclusions are reached.

Newton's second law of motion, which gives rise to classical mechanics, is usually expressed by the following relation between the force and the acceleration vector:

$$\vec{F} = m\vec{a} \quad \Leftrightarrow \quad \vec{F} = m\frac{d\vec{v}}{dt} \qquad (1.1)$$

The proportionality constant m, called mass, represents a measure of the inertia of the rigid body on which the force acts.

Starting from the relation (1.1), the energy transferred to the mass m by the force F along the infinitesimal path element ds can be expressed by the following differential equation:

$$Fds = m\frac{dv}{dt}ds = mvdv \qquad (1.2)$$

Later we will see that the integration of the relation (1.2) results in the kinetic energy of the mass point.

It should be noted that the equations (1.1) and (1.2) presuppose the direct proportionality between force and acceleration of a point mass.

Consequence of this approach is, at changes of speeds, inertia of mass point, on which force affects, is unchanged.

If the fictitious mass point is replaced by real atomic particles, the assumption of the immutability of inertia in the classical view means that, for example, an electron can be accelerated by a strong electric field up to an arbitrarily high velocity.

However, this hypothesis turns out to be wrong for speeds approaching to the speed of light, as experimental observations in particle accelerators show.

The experiments show that an increase in particle inertia can be detected with increasing speed. This leads to a progressive decrease of the particle acceleration.

For speeds approaching those of light, acceleration even approaches zero.

This means that for high speeds there is no longer a direct proportionality between force and acceleration.

So is the basic law of mechanics wrong?

Not at all!

Newton's Second Law of Motion is generally valid

The second law of motion, as Newton formulated it, is correct.

An old physics book of mechanics aptly states:

"The original formulation of Newton's second law of motion is as follows: the force acting on a body is equal to the time derivative of the momentum vector."[3]

---

[3] Daniele Sette - Lezioni di fisica - Volume I, page 100, Nov. 1963. It would be possible to discuss what Newton meant exactly. Important for this work is the correct interpretation of his law.

In his famous work "Philosophiae Naturalis Principia Mathematica" Newton writes:

"Mutation motus proportional esse vi motrici impressae, et fieri secundum lineam rectam qua vis illa imprimitur."

So, *"Mutationem motus"* ... and not *"Mutationem velocitatis"*.

This can be interpreted as follows:

$$\vec{F} = \frac{d\vec{p}}{dt} \quad \Leftrightarrow \quad \vec{F} = \frac{d(m\vec{v})}{dt} \qquad (1.3)$$

According to the correct interpretation of Newton's law, relation (1.3) is the generally valid expression of the second law of motion, and therefore, as we will show later, it should be used instead of relation (1.1) for a correct treatment of mechanics in the most general case.

In equation (1.3) $\vec{p}$ represents the momentum vector of the rigid body. It is identical to the product $m\vec{v}$ of mass and velocity.

It is obvious that (1.3) can be reduced to (1.1) if the mass remains nearly unchanged, which is certainly valid for $v \ll c$.

However, for velocities approaching the speed of light, mass must be considered as a function of velocity, therefore its differentiation is required.

Fundamental Importance Relationship

If the elementary work is expressed by the force F along the infinitesimal path element ds, starting from equation (1.3) and analogously to relation (1.2), the following differential equation[4] results:

---

[4] From now on, the formulas dispensed with the use of vectors, and instead their amounts are used in the assumption that the path ds is always parallel to the force F.

$$Fds = \frac{ds}{dt}dp \quad \Leftrightarrow \quad Fds = vd(mv) \qquad (1.4)$$

Respectively:

$$dE = Fds = mvdv + v^2dm \qquad (1.5)$$

The relation (1.5) is of fundamental importance for the objective of the present work.

It should be noted that equation (1.5) is simplified to (1.2) if the mass, and thus the inertia of the body, remains unchanged ($dm = 0$).

In contrast to relation (1.2), relationship (1.5) takes into account that a transfer of energy to a rigid body is accompanied by an increase in its inertia, as evidenced by experimental observations at high speeds.

Thus, equation (1.5) provides the fundamental energy relation to the further development of Newtonian mechanics for any velocities.

In the course of this work, we will therefore refer to relation (1.5) for the following alternative derivations[5]:

- the relativistic formula of the velocity dependence of mass (Chapter 5)

- the kinetic and the total energy of the rigid body (Chapter 7)

- the relationship of relativistic acceleration depending on speed (Chapter 17).

From relation (1.5) all other relativistic formulas considered in this paper can also be derived indirectly.

---

[5] Appendix AV lists the equations of mechanics discussed in this paper with special reference to the relativistic relations that are directly derived from Newton's second law of motion.

Since the relation (1.5) contains three unequal differentials (ds, dm, dv), it can only be integrated if a second relation between energy and velocity or between mass and energy is known.

With such a relation it would then be possible to eliminate one of the three differentials from (1.5) and thus be able to carry out the integration.

As we will see later, the necessary relation is that of equivalence between mass and energy $E = mc^2$, which will be demonstrated in the third Chapter using classical physics.

Then, the integration of relation (1.5) can be done in the following chapters.

The most commonly used relation of the second law of motion is based on the direct proportionality between force and acceleration. In this form, Newton's law is not suitable for describing physical relationships at high velocities where an increase in body inertia is observed. However, correctly interpreted as a temporal derivation of the momentum, the second law of motion gains the character of a universally valid law and thus provides the suitable expression that takes mass change into account. However, the resulting differential equation cannot be solved without the second relation between energy and mass.

2 Newton's Law - An Analysis

Before starting the first alternative derivation, it is worthwhile to analyze Newton's law in more detail.

As reviewed in the previous Chapter, Newton speaks of "*Mutationem motus*" which is interpreted as variation of momentum. Newton thus bequeathed his law to mankind in the following compact form:

$$\vec{F} = \frac{d\vec{p}}{dt} \qquad \Leftrightarrow \qquad \vec{F} = \frac{d(m\vec{v})}{dt} \qquad (1.3)$$

Or rather, for the derivative with the product rule:

$$\vec{F} = m\frac{d\vec{v}}{dt} + \vec{v}\frac{dm}{dt} \qquad (2.1)$$

You can see that the force vector consists of two terms.

In the first component we find the differential quotient of the velocity which we know as a variable vector, and nothing else is objected first.

In the second component the differential quotient of the mass occurs, and this is remarkable, since in this form the law admits the hypothetical assumption of a variable mass.

Interpreted this way, Newton's law was far ahead of its time, since in Newton's time natural phenomena involving a change in mass were unknown.

In its orbit around the sun the earth reaches a speed of about 30 km/s. This results in a ratio earth / light speed $\beta=0.0001$. The aberration constant of the planet Mercury is somewhat larger with $\beta=0.00016$. And this was probably also the highest speed Newton knew, if at all.

Today we know that at these speeds, due to the kinetic dependence of inertia, mass change is not noticeable.

Newton was thus unable to verify experimentally a possible dependence of mass on velocity, let alone determine this dependence analytically.

According to Eq. (2.1) he had two terms ($m\frac{d\vec{v}}{dt}$ and $\vec{v}\frac{dm}{dt}$) at his disposal, with which he could have explored his mechanics, but according to the knowledge of his time he used only the first one:

$$\vec{F} = m\frac{d\vec{v}}{dt} \qquad (2.2)$$

From this evolved classical mechanics which has always been a borderline case in physics and still remains so today.

So, Newton was not necessarily an optimal user of his own laws. And his idea of space and time is not correct from today's point of view.

So postulates Newton on time: *"Absolute, true, and mathematical time, of itself, and from its own nature, flows equably without relation to anything external..."*

And for the space he claims: *"Absolute space, in its own nature, without relation to anything external, remains always similar and immovable."*

In this regard, Newton was not right.

Now, however, we want to see what conclusions we can come to today on the basis of experiments that were impractical in Newton's time.

For this purpose, Relation (2.1) is used for a more detailed investigation in energy form.

We introduce the scalar product of the force vector $\vec{F}$ with an infinitesimal distance $d\vec{s}$ running in the direction of action of the force. An infinitesimal work is performed, transferring an infinitesimal energy:

$$\vec{F} \cdot d\vec{s} = F ds = dE$$

From (2.1) we get:

$$dE = m\frac{ds}{dt}dv + v\frac{ds}{dt}dm$$

Or:

$$dE = mvdv + v^2dm \qquad (2.3)$$

In (2.3), the two terms that have now been redesigned are still clearly recognizable:

- The first one, $mvdv$, describes the speed change,
- the second one, v^2dm, the mass change.

Consequently, an infinitesimal energy input dE to linear motion generally causes not only a change in velocity, but also a possible change in mass.

Considerations on the Speed

Now the differential equation (2.3) should be integrated. For this the relation is necessary which describes the dependence of the mass on the speed. However, this relation was unknown in Newton's time.

Today, physicists have access to particle accelerators that can be used to perform experiments at different velocities.

Case $v \ll c$**:** The experiments show that at low speeds the mass of the particles remains practically constant $(dm/dt \approx 0)$. In these cases, the infinitesimal energy dE supplied to the body only affects the first term of equation (2.3), and therefore this relation can be simplified as follows:

$$dE = mvdv + \cancel{v^2 dm} \qquad (2.4)$$

The differential equation (2.4) can be easily integrated to deduce the relation of the kinetic energy:

$$E = \frac{1}{2}mv^2$$

And so can continue to build the classical mechanics on it.

Case $v \rightarrow c$**:** At remarkably high speeds close to c, however, something completely different results.

The experiments show that the particles can hardly be accelerated near this speed. So, their mass seems to grow while their velocity remains nearly constant at values just below c. It follows $dv/dt \approx 0$ even with not excessively large values of the mass. Thus, the term $mvdv$ becomes negligibly small compared to $v^2 dm$.

At very high speeds near c (2.3) becomes (2.5):

$$dE = \cancel{mvdv} + v^2 dm \qquad (2.5)$$

This shows that in this case the increase in energy practically only affects the second term of relation (2.3). This causes the mass of the system to increase at an almost constant speed.

From (2.5) for $v \rightarrow c$ the following equation can then be derived:

$$dE = c^2 dm \qquad (2.6)$$

... which, in infinitesimal form, corresponds to the principle of equivalence of mass and energy $\Delta E = \Delta m c^2$. However, this only applies to $v \approx c$ in this example and is not a general proof of the equivalence principle.

And what can one conclude from this?

The experimental results and the first term of Newton's law give a correct description of the natural phenomena at low speeds, as it is still the case in classical mechanics.

However, the experiments also show that the body's inertia does not remain constant at high speeds, with the following consequences:

- The mass must not be viewed as a constant.
- There is an upper limit on the speed.
- The Galileo transformation is useless for high speeds.
- The second law of motion must be used with the two terms that result from its differentiation, so that it remains generally valid for any speed.

With this last knowledge we can ask ourselves the question:

If only classical mechanics can be derived from the first term of the relation (2.1), what can be achieved by using both terms?

And Newton would certainly have asked himself this question if the results of experiments at high speeds had been available to him.

A concrete answer to this question can be considered as the main task of the present work.

Look back into the past:

At that time Newton did not have any experimental results at high speeds available, so he assumed that the mass remains constant at any speed. Consequently, he used only the first term of relation (2.1).

The physicists, his successors, inherited both terms from him but used only one, even after the results of experiments at high speeds became available to them. And so, the conviction continued to prevail that Newton's law applies only at low speeds and immutable mass.

To this day, this attitude to Newtonian mechanics has not changed.

A Controversial Issue

Wikipedia is a reliable indicator of scientific opinion.

In the English version of the article on Newton's laws of motion, it is asserted that the relation of the second law of motion $\vec{F} = d(m\vec{v})/dt$ holds only for constant mass:

"[...] Since Newton's second law is valid only for constant-mass systems, m can be taken outside the differentiation operator by the constant factor rule in differentiation.[...] " (Status: Nov. 2018).

Other physicists, however, believe that although Newton's second law of motion was conceived for constant mass only, it can be corrected by a *"relativistic intervention"*.

For example, Richard Feynman writes on Newton's second law of motion in his work "Lectures on Physics" (Chapter 15):

„For over 200 years the equations of motion enunciated by Newton were believed to describe nature correctly, and the first time that an error in these laws was discovered, the way to correct it was also discovered. Both the error and its correction were discovered by Einstein in 1905.
Newton's Second Law, which we have expressed by the equation
$$F = d(mv)/dt,$$
was stated with the tacit assumption that m is a constant, but we now know that this is not true, and that the mass of a body increases with velocity. In Einstein's corrected formula m has the value

$$m = \frac{m_0}{\sqrt{1 - v^2/c^2}}$$

where the rest mass m_0 represents the mass of a body that is not moving and c is the speed of light [...]".

In fact, when in equation $F = d(mv)/dt$ the mass m is replaced by the formula $m_0/\sqrt{1 - v^2/c^2}$ of the relativistic velocity-dependent mass, after differentiating we get the expression of the relativistic acceleration (see A II in the Appendix).

This confutes the opinion of those who claim that Newton's law is applicable only to immutable mass.

But that is not all. In the course of this work, it will be shown that the second law of motion remains generally correct, even without the "relativistic intervention" mentioned above. Because by Newton's law without relativistic assumptions exactly the speed-dependent formula $m = m_0/\sqrt{1 - v^2/c^2}$ can be derived for mass without relativistic assumptions (see Chapter 5).

Let us now return to differential equation (2.3)

We have analyzed the two limiting cases where only one of the two terms of (2.3) is appropriate to describe the physical events sufficiently accurately. These are the cases: $v << c$ and $v \to c$.

But what if the speed falls between these two limits, e.g., about in the middle?

In that case, both terms of relation (2.3) must be used to describe the mechanics. Thus, both velocity and mass changes are considered in one formula, and we will see that Newton's law remains valid.

This is essentially "the intuitive approach to relativistic mechanics" that the author of this book wants to demonstrate to the reader.

To avoid misunderstandings in this work, we will use the following names for the physics sections and their results:

- *"Classical mechanics"* refers to the part of mechanics that only works with the first term of relation (2.3).

- *"Newtonian mechanics"* is the part of mechanics that uses both terms of relation (2.3). We will see that, starting from it, the formulas of Theory of Special Relativity can be derived alternatively.

- *"Classical physics"* are those areas of physics that do without the concepts of quantum mechanics and without the direct or indirect use of Lorentz transformations.

- *"Relativistic derivations"* refers to the proofs of formulas which directly or indirectly use the Lorentz transformations.

- *"Non-relativistic* or *alternative"* derivations are the proofs that do not require direct or indirect use of the Lorentz transformations.

- The results of *"classical physics"* are all formulas and methods that can be theoretically proven for significantly lower velocities than those of light and / or can be experimentally verified for $v \ll c$. This includes the formulas of energy, momentum, and Doppler Effect for electromagnetic waves at low speeds.

These results (and only them) are used in this work to derive the formulas of the Theory of Special Relativity alternatively.

3 Proofs of $E = mc^2$ from classical physics

The demonstration of the equivalence between mass and energy $E = mc^2$, made in the context of the traditional interpretation of the Theory of Relativity, appears at the end of a long chain of derivations.[6]

The beginning of the chain is given by the Lorentz transformation on which the theory itself is based.

Because it is considered by many to be the most important result of relativity, the mass-energy equivalence has always been the most convincing argument in favor of the theory for anyone reluctant to accept its paradoxical assumptions.

Indeed, how can one acknowledge the validity of such an important formula without accepting the theory from which it arose?

On the other hand, it is an obvious shortcoming of the method to try to persuade critics with results rather than with convincing arguments.

Nevertheless, for many physicists and advocates, who see $E=mc^2$ as the best confirmation of Einstein's theory, the relativistic origin of this equation has become almost a dogma.

It is therefore irritating for them to be confronted with the assumption that the equivalence of mass and energy does not necessarily presuppose the theory of relativity.

---

[6] In Max Born's seminal work on relativity, "Die Relativitätstheorie Einsteins," a demonstration of $E = mc^2$ is made by approximation using the relativistic mass formula, after deriving the latter from the relativistic composition of velocities, which in turn is derived from the Lorentz transformation. Commenting on the same demonstration, the authors Brandes and Czerniawski in their work "Spezielle und Allgemeine Relativitätstheorie" state about $E = mc^2$: "Eine exakte Herleitung gibt es nicht" translated, "There is no exact derivation." As the reader can easily ascertain, this statement is refuted in this chapter.

$E = mc^2$ as a Proof of the Validity of the Theory of Relativity

In the course of this discussion, we will ascertain that the equivalence between mass and energy, derived from classical physics, is indeed the best proof of the validity of the Theory of Relativity, however, not as a result, but in the role of a prerequisite for an innovative and simpler interpretation of the theory itself.

Moreover, irritation and skepticism about the possibility of proving this famous equation even with classical physics are unfounded, since Einstein himself provided convincing proof of this possibility.

In his work "Die Relativitätstheorie Einsteins" (fifth edition, Springer-Verlag, p. 244) Max Born states:

Einstein's equation $E = mc^2$, which determines the proportionality of energy and inertial mass, has often been called the most important result of the Relativity Theory. Therefore, let us give <u>another simple proof that comes from Einstein himself and makes no use of the mathematical formalism of the Theory of Relativity</u> [Emphasis added by the author]. This relies on the fact of the existence of the radiation pressure. That a wave of light which appears on an absorbing body exerts a pressure on it follows from Maxwell's field equations with the aid of a theorem first derived by Poynting (1884). According to this theorem, it turns out that the momentum exerted on the absorbing surface by a short flash or hit of light, having energy E, is equal to E / c. ...".

$E = mc^2$ is not an Exclusive Result of the Theory of Relativity

The derivation that follows this quote proves, contrary to the general belief, that the principle of equivalence of mass and energy is not a compelling result of the theory of relativity, because the derivation of the equation $E = mc^2$ cited by Max Born is based on a theorem from the year 1884.

At that time there was only classical physics. The Theory of Relativity and Quantum Mechanics were developed later.

In the following, I would like to report my own demonstration of the mass-energy equivalence principle $E = mc^2$, which is similar to the one mentioned above and is based on the same phenomenon of the so-called *"radiation pressure"*.

Derivation of $E = mc^2$ based on the "radiation pressure".

The thought experiment considered here uses, instead of the effect of the radiation emitted in a tube, as described by Max Born, the observation of the phenomenon of emission and absorption of a photon between two physical bodies.

We consider a physical system consisting of two identical bodies K1 and K2 of mass m, which initially rest at a distance l from each other.

It is assumed that the bodies do not exchange any energy or matter with the environment. It is also assumed that no external forces act on the bodies.

Because of these assumptions, the following conditions apply to any change in the internal state of the system:

1. **The total mass of the system remains unchanged**

2. **The center of mass of the system remains at rest**

With the same masses, the center of gravity B of the system lies exactly midway between the bodies at a distance $l/2$ to both, as shown in Figure 1.

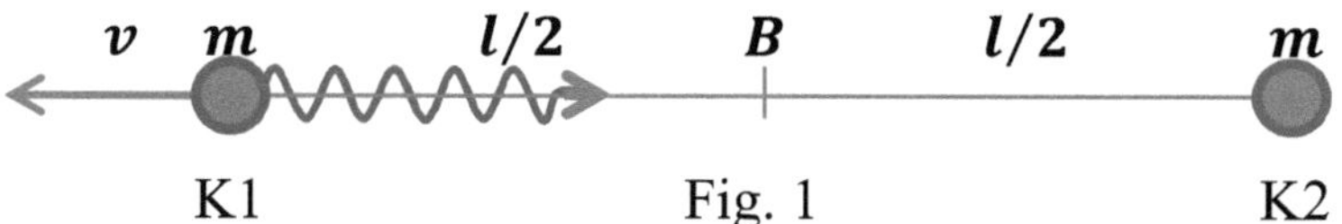

It is assumed that at a certain point in time the left body K1 emits a short intense beam of light in the direction of the right body K2.

As a result of the emission of the light beam, the emitting body K1 receives a recoil and then moves in the opposite direction to the light beam at the velocity *v*.

The First Relationship concerns Time

We assume that after the time *Δt* from the emission, the light beam has reached the body K2 and is absorbed by it. During the same time interval, the body K1 has traveled the distance *Δl*. Therefore, the following relationship applies to the time interval (*c* = speed of light):

$$\Delta t = \frac{l}{c} = \frac{\Delta l}{v}$$

It follows:

$$v = \frac{\Delta l}{l} c \qquad\qquad (3.1)$$

Figure 2 shows the situation at the moment the light beam is absorbed by the right body.

According to the assumption, no external forces act on the system, so the position of body K2 has not changed in the meantime.

The body K1 on the left has moved away during the period *Δt* and is now *l* + *Δl* from the body K2.

After a cursory observation, one might think that the center of gravity *B* of the system would also have shifted. However, this is not possible because, according to the prerequisite, no external forces act on the system.

The Second Relationship concerns Mass

From this it can be concluded that, because of the emission of the light beam, the body K1 on the left, which is located at the greater distance from the center of gravity, must have decreased by a certain mass Δm that has yet to be calculated.

On the other hand, since the total mass of the system remains unchanged, the mass of the body K2 on the right must necessarily have increased by the same amount Δm.

It follows that after the absorption of the light beam the mass of the body K2 has become $m + \Delta m$ and the mass of the body K1 has become $m - \Delta m$.

K1 Δl $l/2$ B $l/2$ K2

$m - \Delta m$ Fig. 2 $m + \Delta m$

The following applies to the position of the center of gravity between two masses as in Fig. 2:

$$(m - \Delta m)(\Delta l + l/2) = (m + \Delta m)\, l/2 \quad \Rightarrow$$

$$m\Delta l - \Delta m\Delta l + m\, l/2 - \Delta m\, l/2 = m\, l/2 + \Delta m\, l/2 \quad \Rightarrow$$

$$\frac{(m - \Delta m)\Delta l}{l} = \Delta m \qquad\qquad (3.2)$$

With the relation (3.2) we now have the second equation necessary to produce the proof.

The third equation required to derive the principle of equivalence is provided by Poynting's theorem from 1884, mentioned at the beginning of this Chapter.

From this theorem it follows that the momentum p of a light beam with energy E is: $p = E/c$. Where c is the speed of light.

The Third Relationship concerns Momentum

The law of conservation of momentum applied to the process of light emission shows that the counter impulse received from the left body K1 must be equal to the momentum of the light beam:

$$(m - \Delta m)v = \frac{E}{c} \tag{3.3}$$

Using relation (3.1) in equation (3.3) gives:

$$(m - \Delta m)\frac{\Delta l}{l}c = \frac{E}{c} \quad \Rightarrow \quad (m - \Delta m)\frac{\Delta l}{l} = \frac{E}{c^2} \tag{3.4}$$

And taking into account relation (3.2) we then get:

$$\Delta m = \frac{E}{c^2}$$

It can therefore be concluded that:

- The emission of a light beam with the energy E by a body causes a decrease of the mass of the body itself, which is equal to the energy of the light beam divided by the square of the speed of light.

- The absorption of a light beam with the energy E by a body causes an increase in the mass of the body itself, equal to the energy of the light beam divided by the square of the speed of light.

Given the simplicity of the demonstration just made, it is logical to ask:

Why persist in wanting to derive this formula in a complicated way, resorting to paradoxical assumptions? Aren't simple solutions also the best?

Derivation of $E = mc^2$ based on the Doppler effect

Another non-relativistic proof of the equivalence principle of mass and energy is based on the Doppler effect of electromagnetic radiation, as already mentioned in the introduction.

The Doppler Effect is treated in "classical" electrodynamics and does not represent a relativistic effect for low speeds of the emitting light source.

To begin with, we would like to refer to classical considerations:

The energy of a quantum of light, also called a photon, is equal to the product hf, and the momentum of a photon is represented by the quotient hf / c where f is the frequency assigned to the photon, h is Planck's constant and c is the speed of light in a vacuum.

It follows that between energy E_f and momentum p_f of a photon the following relationship exists: $E_f = p_f c$.

We consider a body of mass m_1 moving relative to an observer with a low velocity $v_1 \ll c$.

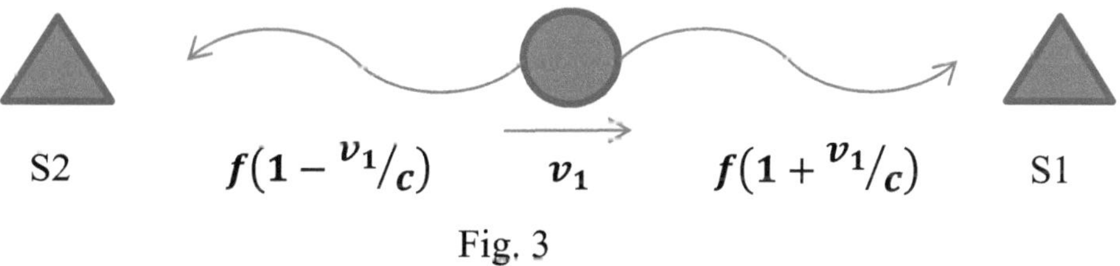

Fig. 3

We assume that at one point in time the body emits two photons of the same frequency: one photon in the direction of the movement, the other one in the opposite direction.

The energy emitted by the body is then: $E = 2hf$.

According to Fig. 3, a spectrometer S1 will measure a frequency $f(1 + v_1/c)$ for the photon in the direction of motion because of the optical Doppler Effect. A spectrometer S2 will measure a frequency $f(1 - v_1/c)$ for the photon sent in the opposite direction.

According to the law of conservation of momentum, it follows that the momentum of the body before emission must be equal to the sum of the momenta of the body and the two photons after emission.

Therefore, we get from the point of view of the observer of the whole system[7]:

$$m_1 v_1 = m_2 v_2 + \frac{hf}{c}\left(1 + \frac{v_1}{c}\right) - \frac{hf}{c}\left(1 - \frac{v_1}{c}\right) \qquad (3.5)$$

Here, the left-directed momentum is negative. It follows:

$$m_1 v_1 - m_2 v_2 = 2\frac{hf v_1}{c^2} \qquad (3.6)$$

where m_2 and v_2 represent mass and velocity of the body after the emission.

---

[7] The derivation is made here using the modern relations $f' = f(1 + v/c)$ and $f' = f(1 - v/c)$ for the optical Doppler effect. Instead, the relations of the Doppler-Fizeau effect from 1848 can also be used. At that time, Fizeau still distinguished between the movement of the light source and the receiver, based on the acoustic Doppler effect. He provided the expressions $f' = f/(1 - v/c)$ for the approach and $f' = f/(1 + v/c)$ for the removal of the source. Using these relations instead of the expressions mentioned above, equation (3.5) must be changed as follows:

$$m_1 v_1 = m_2 v_2 + \frac{hf}{c\left(1 - \frac{v_1}{c}\right)} - \frac{hf}{c\left(1 + \frac{v_1}{c}\right)}$$

Since this equation comes from classical physics, it can only be used for $v_1 \ll c$. The range of validity must therefore only be restricted to values of v_1, for which the result is that v_1^2/c^2 is negligibly small. It follows:

$$m_1 v_1 - m_2 v_2 = \frac{2hf v_1}{c^2\left(1 - \frac{v_1^2}{c^2}\right)}$$

Since in this last relation v_1^2/c^2 can be set to zero without any loss of value, it is reduced to relation (3.6). This proves that the derivation of the equivalence principle E-M can also be carried out with the relations for the Doppler effect by Fizeau from 1848.

Because of the symmetric emission (two equal photons in opposite directions) there will be no change in the velocity of the body after emission. Therefore $v_1 = v_2$ is valid.

On the other hand, the mass of the body does not remain unchanged, otherwise the term on the right side of (3.6) would be zero. But this could only be the case if, contrary to the assumption for the thought experiment, the frequency f or the velocity v_1 would be zero.

Therefore, by substituting v_2 for v_1 and $m_1 - m_2$ for Δm in (3.6), the following equation is obtained.

$$\Delta m v_1 = 2\frac{hf v_1}{c^2}$$

After v_1 is eliminated by division on both sides of this equation and considering that $2hf$ is the energy radiated by the emission of the photons, the formula is given which describes the equivalence of mass and energy in the specific case of electromagnetic emission:

$$\Delta m = \frac{E}{c^2} \qquad \Leftrightarrow \qquad E = \Delta m c^2 \qquad\qquad (3.7)$$

That means:

The radiated energy of a body is equal to the product of its mass decrease and the square of the speed of light.

The alternative derivations of Einstein and the physicist Fritz Rohrlich are based on the principle of conservation of momentum and on the interaction between matter and electromagnetic radiation. Without making use of the mathematical formalism of relativity, they confirm the equivalence of mass and energy in the special case of electromagnetic emission.

4 The Equivalence of Mass and Energy

The derivations of the previous Chapter describe an important aspect of the transformation of mass into energy. Nevertheless, they do not constitute complete proof of the equivalence principle of mass and energy.

In fact, the way in which the relation (3.7) was obtained proves that in electromagnetic emission a mass fraction of a body can be converted into energy.

However, the equation (3.7) does not prove that the entire mass of a body can be converted into energy, nor that conversion of mass into other energy forms besides electromagnetic energy is possible.

Therefore, relation (3.7) does not confirm the equivalence of mass and energy in general.

Electron-Positron Annihilation

The purpose of this Chapter is to fill this gap by considering the experimental observation of "electron-positron annihilation".

This natural phenomenon can be reproduced in special particle accelerators called storage rings.

It is the reaction that can occur through the collision of the electron with the positron, the electron's antiparticle of equal mass and opposite charge:

$$e^- + e^+ \rightarrow 2\gamma$$

As a result of the collision, an unstable particle may form for a very short time.

The decay of this particle can then generate two photons that are emitted in opposite directions.

Fig. 4 illustrates the three phases of the physical process just described:

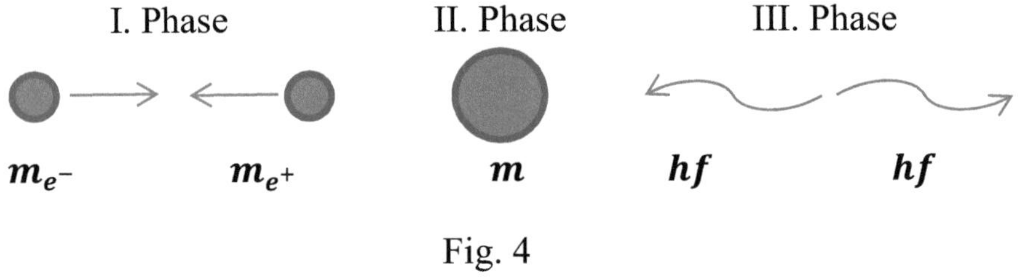

Fig. 4

This phenomenon is similar to the thought experiment described in the third Chapter. The main difference, however, is that in this case, not just a part, but the entire mass of a particle is converted into energy.

We now assume an observer moving at velocity $v \ll c$ relative to the particle formed by the collision of electron and positron. We also assume that the direction of motion of the observer is the same as that of one of the two photons.

By applying the law of conservation of momentum before and after annihilation (phases II and III), from the observer's point of view, considering the optical Doppler Effect, the following equation can be established[8]:

$$mv = \frac{hf}{c}\left(1 + \frac{v}{c}\right) - \frac{hf}{c}\left(1 - \frac{v}{c}\right)$$

This can be simplified as follows:

$$mv = 2\frac{hfv}{c^2}$$

Since $2hf$ is equal to the radiated energy E, we get:

---

[8] Again, the Doppler-Fizeau effect relations can be used alternatively. In this regard, see the note at the bottom of the previous chapter.

$$m = \frac{E}{c^2} \quad \Leftrightarrow \quad E = mc^2 \qquad (4.1)$$

It should be noted that m in relation (4.1), unlike Δm in (3.7), is no longer just a fraction, but rather the entire mass of a particle that has been transformed entirely into energy.

This suggests that each body of mass m can be associate an energy that is expressed by the relation (4.1).

Conservation of Energy in Conversion to Mass

This insight allows us to establish the following energy balance of the three phases of the experiment just described:

If m_e represents the mass of the electron, then (4.1) holds for its internal energy: $E_e = m_e c^2$.

- For the first phase before the collision, the system consisting of the electron-positron pair has a total energy equal to $2m_e c^2 + 2E_k$, where E_k represents the kinetic energy of a single electron.

- Because of the law of conservation of energy, the total energy of the first phase will be transformed into the internal energy mc^2 of the unstable particle formed by the collision (Phase II).

- After the collision, this energy finally passes into the electromagnetic energy of the two emitted photons (Phase III).

For the total energy of the system in the three phases described, the following relationship can be established:

$$E = 2m_e c^2 + 2E_k = mc^2 = 2hf \qquad (4.2)$$

Relation (4.2) confirms the conversion of mass into kinetic as well as electromagnetic energy in a particularly significant case.

It should be noted that both $m_e c^2$ and mc^2 represent only the internal energies of the particles at rest. A relation of the total energy of a particle as a function of the speed has not yet been derived.

Electron-Positron Pair Formation

In addition, it should be pointed out that both the phenomenon of annihilation just described, and the opposite process of so-called electron-positron pair formation occur.

As a result, an electron and a positron are created by the decay of a photon with a minimum energy of $1{,}02\ MeV$.

For higher energies, an increase in the kinetic energy of the particles produced is observed. This confirms the general possibility of converting energy into mass and vice versa of mass into energy.

In this section, the experimental observation "electron-positron annihilation" was considered. Through this physical process, the dissolution of the affected particles and the subsequent emission of two photons can be observed. The analysis of the phenomenon enables us to prove the principle of equivalence mass-energy in the general case. Among other things, the transition of kinetic energy into mass, as well as the complete conversion of the mass of a particle into radiant energy is investigated. These results make it possible to assign an internal energy to a particle at rest according to the equation $E = mc^2$. However, they do not yet provide the formula for the total energy of a point mass as a function of velocity.

5 Dependence of Mass on Velocity[9]

In this Chapter we will see how the formula for the dependence of inertial mass on velocity can be derived from Newton's second law of motion in conjunction with the Mass–Energy Equivalence Principle.

Before making the demonstration, we want to give a conceptual explanation of the dependence of mass on velocity.

Experiments in particle accelerators manifest a phenomenon apparently inexplicable with Newtonian mechanics:

The Inertia of Particles increases as their Velocity increases

In this Chapter we will see that this phenomenon is a consequence of the equivalence principle ($E = mc^2$) between mass and energy.

We remind the reader that the latter principle was demonstrated in the previous Chapters without recourse to relativistic considerations.

By demonstrating the equivalence principle, we have shown that the absorption of radiant energy by a body is accompanied by an increase in the body's mass.

Assuming that E is the absorbed energy, the increase in mass is equal to the quotient E/c^2.

Based on the conservation principle, it is logical to extend this property to all other forms of energy as follows:

---

[9] The following derivation has been prepared by me independently of other physicists in November 2016. It was only after the publication of the first edition of this work that I learned from a reader that a similar derivation was already made in 1961 by Professor Franz von Krbek in his book "Fundamentals of Mechanics".

An absorption of energy results in an increase in the mass of a physical system according to the mass-energy equivalence $E = mc^2$.

This is also what happens when an unconstrained body is subjected to an external force.

In fact, in this case there is an acceleration accompanied by an increase in kinetic energy and, consequently, in mass.

Schematically:

Increase in velocity➔ Increase in kinetic energy ➔ Increase in mass

In quantitative terms this concept translates into the following identity:

Mass of the body in motion = mass of the body at rest + mass of the kinetic energy of the body.

Consequently:

The Inertia of a Body depends on its Kinetic Energy

Therefore, an increase in velocity causes an increase in the mass of the "system" consisting of the body and its kinetic energy.

The dependence of mass on velocity is thus a direct consequence of the principle of equivalence between mass and energy.

With this concept we diverge in this work from conventional demonstrations of the relativistic mass formula.

In fact, the interpretation of Lorentz and Einstein is based on the contraction of the lengths and is therefore difficult to understand.

An increase in mass due to an increase in kinetic energy, on the other hand, is intuitively clear.

This also establishes a direct connection between Newtonian and relativistic mechanics.

That being premised, we can now proceed with the derivation of the relativistic mass formula.

After the demonstration of the equivalence of mass and energy with classical physics, this is the second demonstration of fundamental importance for the purposes of this paper.

Indeed, with the following derivation of relativistic mass we obtain the first relation containing the Lorentz factor.

We thus enter the field of application of the Theory of Relativity starting from Newtonian mechanics, without assuming either the postulate of the constancy of the speed of light or the use of Lorentz transformations.

On the other hand, the relativistic mass formula is the basic relation in the alternative path discussed here.

In fact, it is used to derive all other demonstrations including the theoretical demonstration of the constancy of the speed of light.

Description of the Demonstration

We assume that a constant force F acts on a point mass.

In the first Chapter it was shown that in the most general case, if the path runs in the same direction of the acting force, the infinitesimal work Fds of the force, or rather the contribution of elementary energy dE provided to a body, can be expressed by the following differential equation, which is derived directly from Newton's second law of motion:

$$Fds = dE = mvdv + v^2dm \qquad (1.5)$$

In (1.5), m is a measure of the inertia of the body. From a physical point of view, however, it is not just the mass of the body, but, as we have seen, the mass of the entire "system" consisting of the body and its kinetic energy.

Thus, relation (1.5) shows that an energy input generally causes not only an increase in the velocity of the point mass ($mvdv$), but also an increase of its inertia (v^2dm).

If the speed is considerably lower than the speed of light, then only the contribution of the mass of the body is significant for the inertia since the kinetic energy remains low.

At high speeds, the mass of the kinetic energy is no longer negligible. It becomes dominant at speeds close to the speed of light and thus inhibits the acceleration of the particles.

If we assume that a constant electric force acts on a progressively increasing mass, then it should be understandable why a particle can only be accelerated to a certain speed and not further.

It should be noted that this last point is particularly important for an intuitive understanding of the process that will lead to the derivation of the first relativistic equation in this Chapter.

In the first Chapter it was stated that the equation (1.5) cannot be solved by integration, unless some other relation between energy and mass is known.

The derivations made in the previous Chapters 3 and 4 fill this gap by providing the missing relation, because from the equivalence equation of mass and energy $E = \Delta mc^2$ it can be concluded that not only mass can transform into energy, but also that every supply of energy is accompanied by an increase of mass.

In other words, *"mass is energy, and energy has mass."*[10]

Based on this, it can be stated that the inertia associated with the energy input dE in (1.5), corresponds to the following mass increase dm:

$$Fds = dE = c^2 dm \qquad (5.1)$$

The substitution of Fds by $c^2 dm$ makes it possible to eliminate the differential ds from equation (1.5). Thus, an integrable differential equation results only as a function of mass and velocity:

$$c^2 dm = mvdv + v^2 dm \qquad (5.2)$$

The result of the integration of (5.2) gives the relation of the dependence of the mass on the velocity.

The relation (5.2) can also be rewritten as follows:

$$\frac{dm}{m} = \frac{v}{c^2 - v^2} dv \qquad (5.3)$$

If the second side of equation (5.3) is integrated between the integration limits 0 and the indefinite value v of the velocity and m_0 denotes the mass corresponding to zero velocity (the so-called rest mass), then:

$$\int_{m_0}^{m} \frac{dm}{m} = \int_{0}^{v} \frac{v}{c^2 - v^2} dv = -\frac{1}{2} \int_{0}^{v} \frac{d(c^2 - v^2)}{c^2 - v^2} \qquad \Rightarrow$$

$$[ln(m)]_{m_0}^{m} = -\frac{1}{2}[\ln(c^2 - v^2)]_{0}^{v} \qquad \Rightarrow$$

---

[10] Albert Einstein, Leopold Infeld – Die Evolution der Physik, page 267 – Weltbild Verlag

$$ln\frac{m}{m_0} = \frac{1}{2}\,ln\frac{c^2}{c^2 - v^2} \qquad\qquad \Rightarrow$$

$$\frac{m}{m_0} = \sqrt{\frac{c^2}{c^2 - v^2}} \qquad\qquad \Rightarrow$$

$$m = \frac{m_0}{\sqrt{1 - \dfrac{v^2}{c^2}}} \qquad\qquad (5.4)$$

The relationship (5.4) expresses the dependence of inertia on velocity for a body of mass m_0.

Thus, one of the most important results of the Theory of Special Relativity is confirmed without the use of the Lorentz transformations, which are the basis of Einstein's theory.

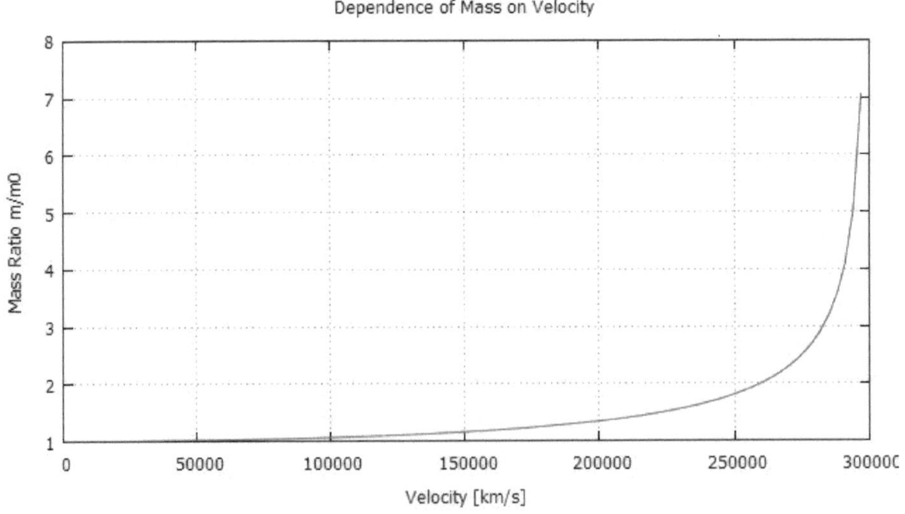

Fig. 5

In his work "Theory of Relativity" Wolfgang Pauli explains:

"This expression of mass dependence on velocity was first derived by Lorentz for the mass of the electron, assuming that also the electrons undergo the Lorentz contraction during motion."[11]

Innovative Approach to the Theory of Relativity

One thing is certain: With the alternative derivation of the dependence of mass on velocity, we have finally left the field of application of classical physics and entered the field of relativity theory. This is particularly evident since the formula in question contains the Lorentz factor.

However, it is noteworthy that in the present work the formula (5.4) was derived starting from classical physics and without assuming the paradoxical hypothesis of length contraction, unlike relativistic proof based on the Lorentz transformation.

Thus, we have achieved the most important result without setting up postulates or resorting to the paradoxical premises of Einstein's established interpretation.

Throughout the remainder of this work, the relationship (5.4) is frequently employed to derive additional relativistic formulae.

Thus, m_0 always denotes the invariant mass (also called rest mass) of a body at zero velocity. Instead, m denotes the total mass that can be assigned to a body as a function of its speed[12].

---

[11] Wolfgang Pauli – Relativitätstheorie, page 97 – Springer Verlag

[12] It should be clarified at this point: To the inertia of the body, two mass fractions contribute (i) on the one hand, the mass m_0 of the body itself, and (ii) on the other hand, the mass that can be assigned to the kinetic energy of the body. The last part can be several orders of magnitude larger than the mass m_0 of the body.

Starting from equation (5.4), when both sides of the equation are multiplied by velocity, the formula for the momentum in the general case is:

$$\vec{p} = \frac{m_0 \vec{v}}{\sqrt{1 - \dfrac{v^2}{c^2}}} \qquad (5.5)$$

Relation (5.5) shows, in agreement with the experimental observations, that a body can neither exceed nor reach the speed of light.

The immediate consequence of this conclusion is the inapplicability of the Galilean transformation at any speed, with the consequence that the help of transformations must be dispensed with in the further course of this treatise.

At this point we would like to anticipate that multiplying both sides of equation (5.4) by the square of the speed of light gives the formula for the total (i.e., internal + kinetic) energy of a body as a function of its velocity:

$$mc^2 = \frac{m_0 c^2}{\sqrt{1 - \dfrac{v^2}{c^2}}} \qquad (5.6)$$

Consider that (5.6) is only anticipated here. Its proof follows in the seventh Chapter.

> The proof of the equivalence principle of mass and energy presented in chapters 3 and 4 makes it possible to assign the corresponding inertia to the energy transferred to a body. Thus, one obtains the necessary relation for solving the differential equation from Newton's second law of motion. Through integration one can derive the relationship between a body's inertia and its speed. This proves one of the most important results of the Theory of Special Relativity without resorting to the Lorentz transformation or presupposing length contraction.

6 Traditional and Alternative Derivation of Relativity

In the previous Chapters, we used an innovative method to derive the equivalence of mass and energy, as well as the relativistic mass formula.

Since these are the fundamental relations that will be used for the alternative derivation of the Theory of Special Relativity, one should now have a concrete idea about the assumptions and tools that will be used in this work.

Therefore, before proceeding with the demonstrations, we will compare the traditional method of deriving the theory with the alternative method proposed here.

It should be noted that, although using different interpretations, the two methods lead to the same results.

A brief description of the two methods follows.

Traditional Derivation Method

The traditional method of derivation is based on the postulate of the constancy of the speed of light. This principle could be verified experimentally, but could not be proved theoretically because it contradicts classical mechanics.

Physical Assumptions of the Traditional Method

The basic assumption is an extension of the principle of relativity, which states that the speed of light remains constant in all inertial frames of reference. This confirms that, even for information transport, there is no preferred inertial frame of reference.

There follows the untenability of the classical addition of velocities and the consequent incompatibility with the Galilean transformation.

Thus, with the derivation of the Lorentz transformations for space and time, a new interpretation of kinematics is outlined, with the following consequences: length contraction, time dilation and other paradoxical assumptions.

Unlike the Galilean transformation, it is recognized that the Lorentz transformation is also consistent with Maxwell's laws of electromagnetism. The relativistic velocity addition formula is derived from this.

Relativistic mechanics is then developed on the basis of the new kinematics.

Within relativistic kinematics, constant mass is no longer consistent with the law of conservation of momentum.

The relativistic mass formula is derived from the velocity addition theorem, confirming that the experimentally observed variation of the inertia of bodies is a consequence of the dependence of their mass on velocity.

All other relativistic relations are also derived by assuming the Lorentz transformation. The latter is the core of the traditional derivation method.

Consequences of the Traditional Interpretation

Space and time can no longer be considered absolute. They lose their invariance property and must be regarded as dependent on the speed of the frame of reference.

Conclusion

The traditional derivation method is based on partly paradoxical premises that conflict with classical mechanics.

There is no direct connection between classical and relativistic mechanics.

The traditional derivation method has caused a revolution in the field of physics. In the past, it has aroused harsh criticism from its opponents and boundless admiration from its supporters. Despite its appeal, however, the method remains difficult to understand and is not intuitive.

Alternative Derivation Method

The alternative derivation method presented here is based on the equivalence principle of mass and energy $E = mc^2$, which is derived from the laws of classical physics.

Newton's second law of motion, using the mass-energy equivalence, leads to the derivation of the relativistic mass formula.

The derivation of this first formula, which includes the Lorentz factor, establishes a logical connection between Newtonian and relativistic mechanics.

Physical Assumptions of the Alternative Method

This procedure makes it easy to recognize that, at high speeds, the dependence of a body's inertia on velocity is due to the mass associated with its kinetic energy and not to the variability of its mass as a function of speed.

Relativistic mechanics is later developed using the equivalence of mass and energy, the relativistic mass formula and the laws of conservation of energy and momentum. Transformations for space and time are not used.

In this way, among others, the relation of the relativistic addition of velocities is also proved.

From this formula it is possible to derive the constancy of the speed of light.

Thus derived, the latter should no longer be considered a postulate, but rather a theoretically provable physical principle.

Using the principles of conservation of energy and momentum, the relationship of the contraction of lengths as a function of velocity is also demonstrated.

From the latter formula, it is possible to derive relativistic transformations for space and time, which turn out to be identical to those obtained by Lorentz from the laws of electromagnetism and by Einstein with the hypothesis of the constancy of speed of light.

Consequences of the Alternative Method

The consequences of the alternative derivation method are the same as those of the traditional derivation method: space and time can no longer be considered absolute. They lose their invariance property and must be regarded as dependent on the speed of the frame of reference.

In the alternative derivation method, however, these consequences do not appearat the beginning, as in the interpretation of Einstein and Lorentz, but at the end of the series of demonstrations of the theory.

Conclusion

The alternative derivation method arrives at the same results as the traditional one, without using either postulates or paradoxical premises.

Starting from the principle of mass-energy equivalence, which can be derived from classical physics, it establishes a direct connection between classical and relativistic mechanics.

The alternative derivation method described here is intuitive and easy to understand.

7 Calculation of Kinetic and Total Energy

In this Chapter we will see how the relativistic formula for kinetic and total energy can be derived from the second law of motion in connection with the equivalence principle of mass and energy.

Before demonstrating it, we want to give a conceptual explanation of the final result.

Heuristic Explanation of the Derivation

It follows from the principle of mass and energy equivalence that the energy corresponding to a mass m_0 is equal to $m_0 c^2$.

In the demonstration of the above principle, it was assumed that the mass m_0 is at rest.

However, this condition is not necessary. It implies the possibility of extending mass-energy equivalence to masses in motion as well.

In the most general case we can assert that the equivalence principle can be expressed by the following relationship:

$$E = \frac{m_0 c^2}{\sqrt{1 - \dfrac{v^2}{c^2}}} \qquad (7.0)$$

Equation (7.0) is obtained by substituting for the mass at rest the corresponding term of a mass in motion as a function of velocity (see Chapter 5).

In the latter case, relation (7.0) expresses the total energy of a body in motion.

Since it is an unbound body, its total energy must necessarily be equal only to the sum of its kinetic and rest energies:

$$\frac{m_0 c^2}{\sqrt{1 - \dfrac{v^2}{c^2}}} = E_k + m_0 c^2 \qquad (7.5)$$

Thus, the following relationship is then derived for the kinetic energy:

$$E_k = \frac{m_0 c^2}{\sqrt{1 - \dfrac{v^2}{c^2}}} - m_0 c^2 \qquad (7.4)$$

Now let us see how Newton's second law of motion can provide us with a rigorous demonstration of the same relationship between the kinetic energy and total energy of the body.

Demonstration with Newton's Law

The second law of motion, in conjunction with the relation $E=mc^2$ and the relativistic mass formula, allows an alternative derivation of the relativistic energy of the body.

Since both the principle of equivalence between mass and energy $E=mc^2$ and the formula for mass as a function of velocity have been proved without the aid of relativistic axioms, this derivation of relativistic energy constitutes the third link in the chain of proofs leading from classical physics to Special Relativity by a simple and easily accessible alternative path.

The relativistic energy formula derived here is later used, together with that of momentum, to prove all the other formulas of Special Relativity, including that of the relativistic addition of velocities.

This latter relation thus makes it possible to prove theoretically the constancy of the speed of light.

Calculation of the Kinetic Energy of Classical Mechanics

The definition of work in mechanics can be used to calculate the kinetic energy supplied to an unbound rigid body by a force acting on it.

In classical mechanics, this is usually done as follows:

Assuming that the mass is constant, the differential equation (1.2) can be applied.

As seen in the first Chapter, equation (1.2) is a direct consequence of equation (1.1):

$$\vec{F} = m_0\vec{a} \quad \Leftrightarrow \quad \vec{F} = m_0\frac{d\vec{v}}{dt} \qquad (1.1)$$

Consequently, for the supply of kinetic energy by the infinitesimal element of the work $\boldsymbol{Fds}$, holds[13]:

$$dE_k = Fds = m_0 vdv \qquad (1.2)$$

If the differential equation (1.2) is integrated for an initial velocity equal to zero, the expression for the kinetic energy results:

$$E_k = m_0 \int_0^v v\,dv = \frac{1}{2}m_0 v^2 \quad (v \ll c) \qquad (7.1)$$

Since it is derived from equation (1.1), relation (7.1) describes the kinetic energy of a point mass m_0 only for velocities well below the speed of light, where the inertia of the body remains practically unchanged.

---

[13] As already mentioned, it is also assumed here that the infinitesimal distance element ds runs in the same direction of the force F.

In the more general case, i.e. even for speeds approaching the speed of light, it is necessary to use the following relationship instead ...

$$dE_k = Fds = v^2 dm + mvdv \qquad (1.5)$$

... in accordance with what was stated in the first Chapter[14].

If the two relationships considered in the fifth Chapter are applied for the mass, ...

$$Fds = dE_k = c^2 dm \qquad (5.1)$$

$$m = \frac{m_0}{\sqrt{1 - \dfrac{v^2}{c^2}}} \qquad (5.4)$$

... then it can be stated that by substitution, the velocity-dependent mass m can be eliminated from the differential equation (1.5).

This results in the required relationship to solve the differential equation (1.5).

From (5.1) follows:

$$dm = \frac{dE_k}{c^2} \qquad (7.2)$$

After substituting (5.4) and (7.2) into (1.5), the following relationship between the supplied mechanical energy and the velocity results:

---

[14] It should be remembered that at this point, as in the remainder of this work, always unbound rigid bodies are considered that have no potential energy. The bodies considered here can therefore be adapted to the subatomic particles. It is therefore obvious that the infinitesimal work in the differential equation (1.5) completely merges into the kinetic energy of the point mass.

$$dE_k = v^2 \frac{dE_k}{c^2} + \frac{m_0}{\sqrt{1 - \dfrac{v^2}{c^2}}} v\,dv \qquad\qquad \Rightarrow$$

$$(1 - \frac{v^2}{c^2})dE_k = \frac{m_0 v}{\sqrt{1 - \dfrac{v^2}{c^2}}}\,dv \qquad\qquad \Rightarrow$$

$$dE_k = \frac{m_0 v}{\left(1 - \dfrac{v^2}{c^2}\right)^{\frac{3}{2}}}\,dv \qquad\qquad (7.3)$$

By comparing equations (1.2) and (7.3), it can be seen that in both relations the differential of the kinetic energy dE_k is given only as a function of the invariant mass m_0 and the velocity v.

However, the relation (7.3) is the general expression for the calculation of the kinetic energy at any velocities.

It is easy to show that (7.3) can be reduced to (1.2) for $v << c.$

Just as equation (1.2) provides by integration the kinetic energy of a constant point mass m_0 at low speeds, the integration of the differential equation (7.3) leads to the calculation of the kinetic energy for the general application at any speeds.

To calculate the kinetic energy, the differential equation (7.3) is integrated between the integration limits 0 and v analogously to (1.2):

$$E_k = m_0 \int_0^v \left(1 - \frac{v^2}{c^2}\right)^{-\frac{3}{2}} v\,dv \qquad\qquad \Rightarrow$$

$$E_k = -\frac{1}{2}m_0c^2 \int_0^v \left(1 - \frac{v^2}{c^2}\right)^{-\frac{3}{2}} d\left(1 - \frac{v^2}{c^2}\right) \quad \Rightarrow$$

$$E_k = -\frac{1}{2}m_0c^2 \left[\frac{\left(1 - \frac{v^2}{c^2}\right)^{-\frac{1}{2}}}{-\frac{1}{2}}\right]_0^v \quad \Rightarrow$$

$$E_k = m_0c^2 \left(\frac{1}{\sqrt{1 - \frac{v^2}{c^2}}} - 1\right) \quad \Rightarrow$$

$$E_k = \frac{m_0c^2}{\sqrt{1 - \frac{v^2}{c^2}}} - m_0c^2 \tag{7.4}$$

Equation (7.4) expresses the kinetic energy of a body as a function of its mass and velocity.

It corresponds to the formula for kinetic energy derived by the traditional relativistic derivation method based on the Lorentz transformation.

If we divide the right part of (7.4) by c^2 we get the mass associated with the kinetic energy: $m_c = m_0/\sqrt{1 - v^2/c^2} - m_0$. This is the mass that inhibits particle acceleration at high speeds (see Chapter 5).

Using the series expansion of Taylor and also by the following algebraic method, it can be shown that the relation (7.4) for $v \ll c$ reduces to the equation (7.1):

For velocities much lower than that of light, the quotient $v^4/4c^4$ is negligible compared to the term v^2/c^2.

Therefore, this quotient can be added to the radical of equation (7.4) without changing its value:

$$E_k = \frac{m_0 c^2}{\sqrt{1 - \dfrac{v^2}{c^2} + \dfrac{v^4}{4c^4}}} - m_0 c^2 \qquad \Rightarrow$$

Since the denominator is now the square root of a binomial formula, we get:

$$E_k = \frac{m_0 c^2}{1 - \dfrac{v^2}{2c^2}} - m_0 c^2 \qquad \Rightarrow$$

$$E_k = \frac{m_0 c^2 - m_0 c^2 + \dfrac{1}{2} m_0 v^2}{1 - \dfrac{v^2}{2c^2}}$$

This equation then reduces to (7.1) for $v \ll c$.

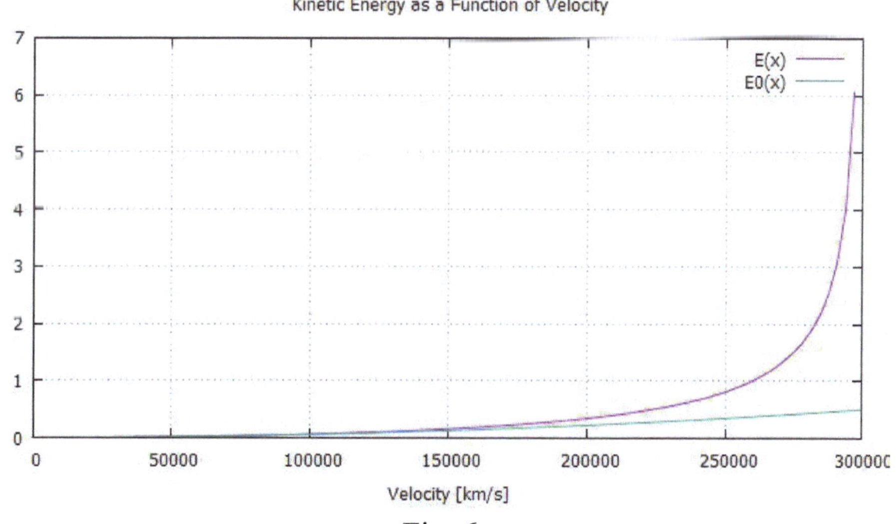

Fig. 6

Fig. 6 compares the kinetic energy curves from equations (7.1) (green curve) and (7.4) (purple curve).

It can be seen that for low velocities the two curves actually coincide.

However, the curves diverge more and more as the velocities approach the speed of light.

Considering the relation (5.4), the formula (7.4) can also be written as follows:

$$mc^2 = \frac{m_0 c^2}{\sqrt{1 - \dfrac{v^2}{c^2}}} = E_k + m_0 c^2 \qquad (7.5)$$

By the relation (7.5) we come to the following important result:

Since the right-hand side of equation (7.5) is equal to the sum of the kinetic and internal energies, we can conclude that mc^2 or $m_0 c^2 / \sqrt{1 - \dfrac{v^2}{c^2}}$ represents the total energy of the point mass m_0 as a function of the velocity.

This confirms what was predicted by Equation (5.6) in Chapter 5, but not demonstrated.

> The relations between the work and the mass as a function of the velocity derived in Chapters 4 and 5 allow substitution to eliminate the velocity dependent mass from the differential equation of the work. The subsequent integration then yields as an end result the expression of the kinetic energy and, at the same time, of the total energy of a point mass as a function of its velocity. These equations also correspond to the relativistic formulas derived by the traditional relativistic derivation method based on the Lorentz transformation.

8   The Relativistic E-p-m Triangle

The relativistic E-p-m triangle illustrates in a truly clear way the relationships that exist between energy, momentum, and mass of the body at high speeds.

In the preceding sections we have seen that in many formulas the reciprocal $\sqrt{1 - \frac{v^2}{c^2}}$ of the so-called Lorentz factor occurs.

This term is reminiscent of the Pythagorean Theorem: According to this, if in a right-angled triangle the hypotenuse is equal to 1 and a cathetus is equal to v/c, then the second cathetus is equal $\sqrt{1 - \frac{v^2}{c^2}}$, as shown in Fig. 7.

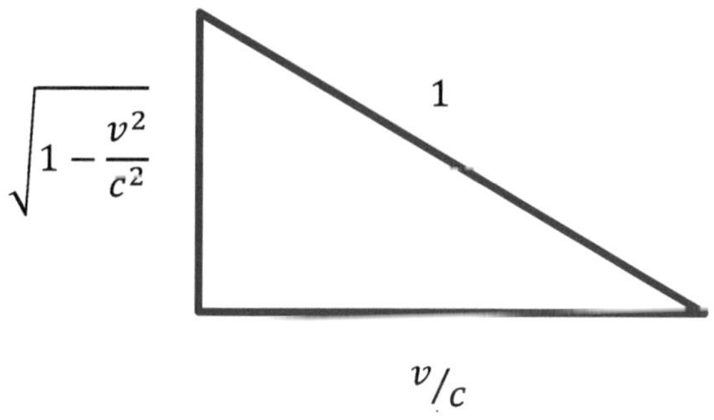

Fig. 7

Now, if all sides of the triangle are multiplied by the total energy mc^2 of a body, the following results are obtained, considering the relation (5.4):

$$\text{1st Cathetus} = mc^2 \sqrt{1 - \frac{v^2}{c^2}} = m_0 c^2$$

$$\text{2nd Cathetus} = mvc$$

$$\text{Hypotenuse} = mc^2$$

Obtaining the so-called relativistic triangle shown in Fig. 8:

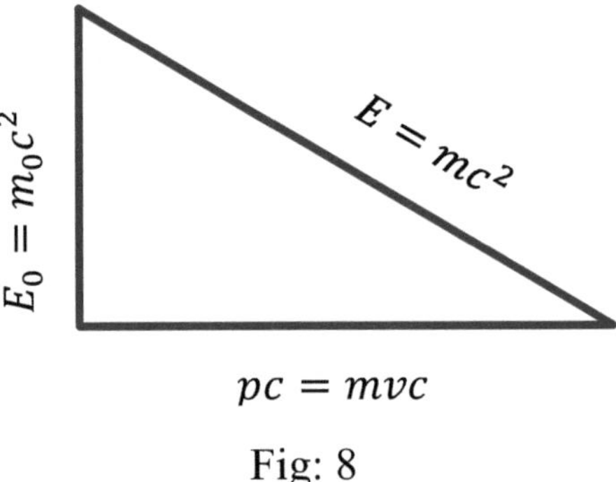

Fig: 8

The following information regarding kinetic energy, total energy, momentum and mass can be derived from the relativistic triangle.

Kinetic Energy

In the previous chapter we showed that relativistic kinetic energy is given by the following relationship:

$$E_k = \frac{m_0 c^2}{\sqrt{1 - \dfrac{v^2}{c^2}}} - m_0 c^2 \qquad (7.4)$$

Or, denoting by m the relativistic mass:

$$E_c = mc^2 - m_0 c^2$$

This relationship can be illustrated geometrically as the difference of the hypotenuse and the vertical cathetus of the relativistic triangle shown in Figure 8.

In general, it can be seen that the longer the vertical cathetus, the lower the kinetic energy and vice versa.

Total Energy

Applying the Pythagorean Theorem to the triangle in Fig. 8 yields the following relations for total energy (E) and momentum (p):

$$E = mc^2 = c\sqrt{p^2 + m_0^2 c^2} \qquad (8.1)$$

Momentum

$$p = \sqrt{\frac{E^2}{c^2} - m_0^2 c^2} \qquad (8.2)$$

By the relation (8.1) it can be seen that when the momentum equals zero, the total energy is reduced to the value $m_0 c^2$. This value corresponds to the internal energy of a point mass m_0.

Mass

Equation (8.2) shows that there can be physical objects with the invariant mass m_0 equal to zero. In this case the momentum is given the value: $p = E\,/\,c$, valid for light quanta[15].

However, equation (8.2) also shows that there cannot be any physical objects with zero energy; otherwise, the momentum would have an imaginary value, which is not allowed.

From the relativistic E-p-m triangle one can also gain another insight:

---

[15] It is interesting to note that at this point the relation of the momentum of electromagnetic radiation reappears, which we used to prove the equivalence principle between mass and energy $E = mc^2$, and which gave rise to the alternative method of proving the theory of relativity described in this paper.

Consequently, if the invariant mass m_0 of a particle equals zero, then the vertical triangle side is equal to zero and the horizontal triangle side becomes equal to the hypotenuse.

It follows:

$$mvc = mc^2 \quad \Rightarrow \quad v = c$$

This means that massless physical objects inevitably move at the speed of light.

Examples of elementary particles that propagate at the speed of light are the exchange particles of the electromagnetic and gravitational interactions. That means the photons and the gravitons.

In the context of Quantum Chromodynamics Theory, it is assumed that the exchange particles of the strong interaction, called gluons, also travel at the speed of light.

The results of the preceding sections can be summarized by the geometrical representation of the so-called relativistic E-p-m triangle. This triangle illustrates in a truly clear way the relationships that exist between energy, momentum, and mass.

9 Velocity Addition of Colliding Electrons

In this chapter, the relativistic addition of velocities is derived with a thought experiment based on the inelastic collision of two electrons.

Two observers O_e and O_L in relative motion examine what is happening independently of one another. Both use the law of conservation of energy for their calculations. Then they compare their results using the rest mass m_0 of the formed particle, which is invariant for both.

Thus, a single relation is formed from two equations, which represents the relative speed between the electrons as a function of the speeds of the single electrons.

With this method we will now derive the relativistic addition formula in the special case of equal velocities as follows.

We imagine the central collision between two unlike charged electrons (electron and positron) moving towards each other with equal velocities v_e.[16]

It is believed that the energy loss caused by the electromagnctic interaction between the electrons can be neglected because of the high speeds.

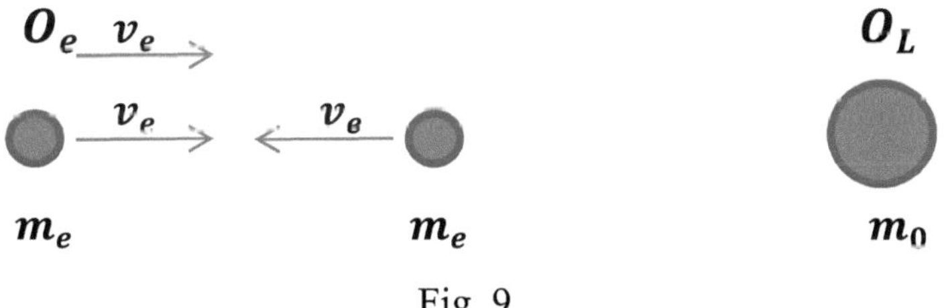

Fig. 9

Suppose that because of the collision, a new particle of mass m_0 is formed. This is from the perspective of the observer O_L at rest.

---

[16] Such an experiment was already carried out in the 1960s in particle accelerators, so-called storage rings, and repeated countless times.

Demonstration using Conservation of Energy

Assuming the observer O_L knows the velocities v_e of the electrons immediately before the collision, then he will be able to calculate the mass of the produced particle using the following equation, which describes the conservation of energy before and after the collision:

$$m_e c^2 + m_e c^2 = m_0 c^2 \qquad (9.1)$$

Where m_e is the mass of the electron in motion as a function of velocity v_e.

That means:

The energy $m_0 c^2$ of the particle produced is equal to the sum of the total energies of the colliding electrons.

Using the relation (5.4), equation (9.1) yields:

$$m_0 = \frac{2m_{0e}}{\sqrt{1 - \dfrac{v_e^2}{c^2}}} \qquad (9.2)$$

Where m_0 and m_{0e} are the invariant masses of the formed particle and the electron, respectively.

Observer O_L can therefore calculate the mass of the produced particle and knows the velocities of the colliding electrons. However, he will not determine that the relative velocity v_{ee} between the electrons is equal to the sum of their velocities $v_e + v_e$, as is evident from the Galilean transformation only for low velocities, because that would not be in accordance with the conservation laws of energy and momentum.

To calculate the relative speed v_{ee} between the two electrons, it is necessary to consider the same thought experiment from the point of view of a second observer O_e who is at rest relative to one of the electrons.

From the point of view of the observer O_e the following total energy E_1 results before the collision:

$$E_1 = m_{0e}c^2 + m_{ee}c^2 = m_{0e}c^2 + \frac{m_{0e}c^2}{\sqrt{1 - \frac{v_{ee}^2}{c^2}}} \qquad (9.3)$$

Where m_{ee} is the mass of the electron in motion as a function of velocity v_{ee}

That is, the energy measured by observer O_e is equal to the sum of the energy of the resting electron and the energy of the electron approaching observer O_e at the velocity v_{ee}.

After the collision, observer O_e, who continues to move at the velocity v_e relative to the formed particle, calculates its energy E_2 using the following formula:

$$E_2 = \frac{m_0 c^2}{\sqrt{1 - \frac{v_e^2}{c^2}}} \qquad (9.4)$$

By replacing the invariant mass m_0 by the expression (9.2) in equation (9.4) and setting $E_1 = E_2$ according to the law of conservation of energy, we get:

$$m_{0e}c^2 + \frac{m_{0e}c^2}{\sqrt{1 - \frac{v_{ee}^2}{c^2}}} = \frac{2m_{0e}c^2}{1 - \frac{v_e^2}{c^2}} \qquad (9.5)$$

Equation (9.5) can be further simplified by dividing all terms by $m_{0e}c^2$:

$$\frac{1}{\sqrt{1 - \frac{v_{ee}^2}{c^2}}} = \frac{2}{1 - \frac{v_e^2}{c^2}} - 1 \qquad \Rightarrow$$

$$\sqrt{1 - \frac{v_{ee}^2}{c^2}} = \frac{1 - \frac{v_e^2}{c^2}}{1 + \frac{v_e^2}{c^2}}$$

From this last expression, the following relation finally results after simple algebraic transformations:

$$v_{ee} = \frac{2v_e}{1 + \frac{v_e^2}{c^2}} \qquad (9.6)$$

Equation (9.6) represents the relative velocity between two particles that are moving towards each other at the same speed, and thus expresses the relativistic addition of velocities at that speed.

Demonstration using Conservation of Momentum

The same result is obtained when the law of conservation of momentum is used instead of the law of conservation of energy.

In this case, for the momentum before and after the collision, we get from the viewpoint of the observer O_e:

$$m_{ee}v_{ee} = \frac{m_0 v_e}{\sqrt{1 - \frac{v_e^2}{c^2}}}$$

Where m_{ee} is the mass of an electron in motion, expressed as a function of its velocity v_{ee}

By using (5.4) and (9.2), we obtain the following result:

$$\frac{m_{0e}v_{ee}}{\sqrt{1 - \frac{v_{ee}^2}{c^2}}} = \frac{2m_{0e}v_e}{1 - \frac{v_e^2}{c^2}}$$

This last equation, solved for v_{ee}, yields relation (9.6) (for details see AIII. in the Appendix).

At this point, we would like to emphasize that the relation (9.6) was derived using only the law of conservation of energy and without invoking the axioms of Special Relativity.

Relation (9.6) shows that, at low velocities, when the term $\frac{v_e^2}{c^2}$ is negligible, the relative speed is given with good accuracy by the sum of the velocities (green Curve in Fig. 10).

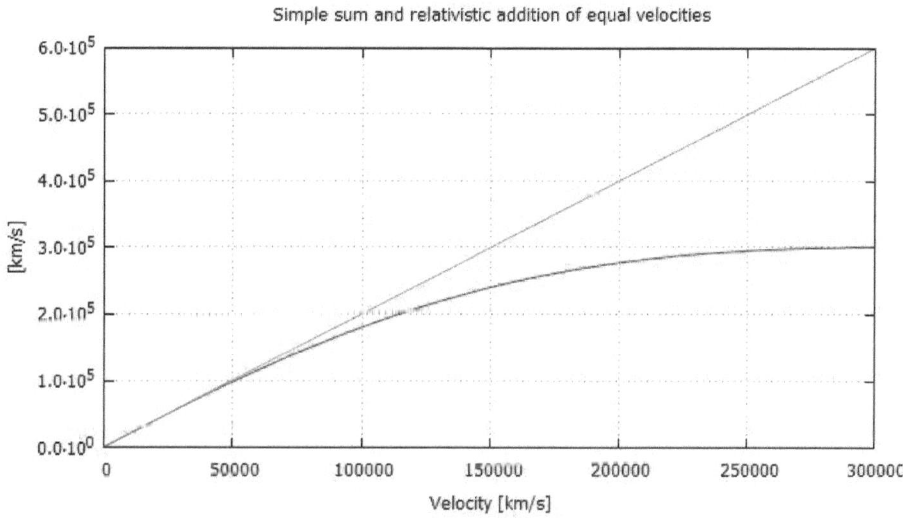

Fig. 10

However, from the relation (9.6) the following general conclusion can be drawn:

High speeds cannot be added together algebraically.

For the addition of equal velocities, the relation (9.6) can be used, which, unlike the simple algebraic addition, complies with the laws of conservation of energy and momentum (violet curve).

It is easy to show that, according to the relation (9.6), the relative velocity v_{ee} between two particles can reach, but not exceed, the speed of light.

The relative velocity v_{ee} reaches the speed of light only if the colliding particles are massless and therefore v_e is equal to c.

If it is assumed that the velocities of the colliding particles are different, an analogous derivation can be carried out for this case, however, this is more complicated algebraically (see Chapter 11).

We will see that, if velocities v_1 and v_2 are not equal, their addition leads to the following result:

$$v_{12} = \frac{v_1 + v_2}{1 + \dfrac{v_1 v_2}{c^2}} \tag{9.7}$$

Relation (9.7) is equivalent to Einstein's relativistic addition formula for velocities and reduces to (9.6) when $v_1 = v_2$

It can therefore be concluded that equation (9.6) represents a first confirmation of Einstein's formula (9.7) for equal velocities.

> The application of the law of conservation of energy to the collision of two electrons represents a first confirmation of Einstein's relativistic addition formula for velocities.

10 Dependence of Time on Velocity

The finding from the previous section that velocities cannot be added algebraically leads to two important consequences.

The first consequence concerns the failure of the Galilean transformation.

When applied to the addition of high speeds, the Galilean transformation leads to incorrect results.

This recognizes the need to define another transformation that remains valid even at high speeds.

The second consequence concerns the time.

What it is about, we will show by the following physical examination.

We will now refer to the thought experiment described in Chapter 8 in order to explore the times measured by two observers in relative motion to each other.

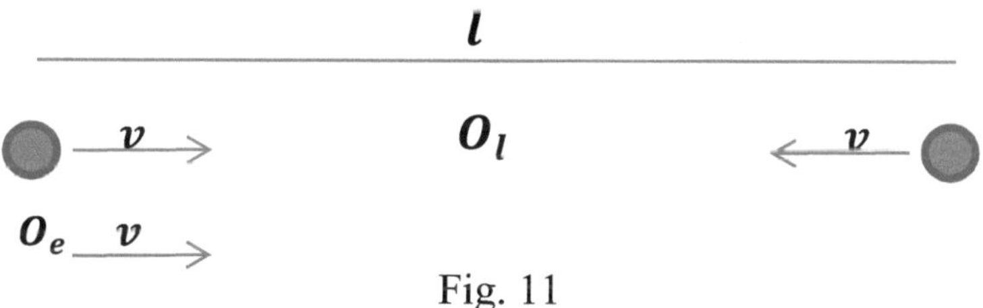

Fig. 11

We assume that in a synchrotron two particles are accelerated towards each other until their velocity v approaches the speed of light.

We imagine that at some point the particles reach the entrance of the accelerator detector, where they move straight up to collision at a constant velocity.

And now we adopt two assumptions, which are self-evident in the context of classical mechanics, but which are not consistent at high speeds, as in the present case, and let's see what happens.

<u>The first assumption is about space</u>. We assume the *invariance of the lengths* for observers in relative motion: We assume that the observers O_l and O_e agree on the length of the detector, i.e., about the distance the particles will travel before the collision.

<u>The second assumption is about time</u>. We assume the *invariance of simultaneity* for observers in relative motion to one another: We assume that from the point of view of both observers, the particles are at the same point in time in the entrance of the particle accelerator detector.

The observers now calculate independently of each other the time that elapses between the entry into the detector and the collision of the particles.

For observer O_l, because he is at rest relative to the experimental laboratory, the particles meet exactly in the middle of the detector and therefore he finds that a single particle travels the path length $l/2$ before the collision.

The times measured by the two observers are discordant

The time measured by him is therefore:

$$t_l = \frac{l}{2v}$$

On the other hand, observer O_e, being at rest at one of the particles (see Fig. 11), calculates the time by dividing the length l of the detector by the velocity at which the second particle approaches him.

We saw in Chapter 9 that this velocity was calculated by the relation (9.6).

The time measured by O_e is therefore:

$$t_e = \frac{l}{2v}\left(1 + \frac{v^2}{c^2}\right)$$

And thus, divergent from the time measured by the observer O_l.

The consequence is that under the given conditions, for observers O_e the particles would no longer meet in the middle of the acceleration detector, just as it has to apply also to him.

This discrepancy could not even be completely eliminated by a transformation leading to a contraction or dilatation of the spatial coordinate, and thus to a change in the detector length.

The incongruence can only be eliminated by postulating that, from O_e observer's point of view, the particles are not at the same time in the entrance of the detector of the accelerator.

Dependence of Simultaneity on the Frame of Reference

This fact indicates a dependence of simultaneity on the choice of the frame of reference.

From this it follows that the correct transformation, not yet derived in this work, must necessarily involve not only the spatial coordinate, but also the temporal coordinate.

Therefore, this implicitly confirms the relativistic postulate of the existence of a local time dependent on the velocity of the frame of reference.

This second consequence is even more serious than the first, which leads to the rejection of the Galilean transformation for the spatial coordinate, because it concerns the untenability of Newton's conception of absolute time.

It should be noted that, at this stage of the study, there is still a lack of the necessary physical knowledge to derive the Lorentz transformations, which provide the correct relations for the temporal and spatial coordinates as a function of the velocity.

One possible way is the following:

From the relativistic velocity-addition formula, the principle of the constancy of the speed of light for any relative speeds between light source and receiver must first be proven. Only then the physicist does have the necessary prerequisites for the derivation of the transformations.

One would get the following results:

After the Lorentz transformations observer O_e perceives a time dilatation[17].

For the length of the detector observer O_e instead establishes a contraction represented by the relationship:

$$l_e = l \sqrt{1 - \frac{v^2}{c^2}}$$

The same length contraction in the direction of motion should also be perceived by the observer O_l for the colliding particles.

For example, protons should no longer be spherical, but like to an ellipsoid.

---

[17] A simple derivation of time dilation can be found in the "thought experiment of the light clock" by Gilbert Newton Lewis and Richard C. Tolman.

For velocities, which are given here as fractions of the speed of light, the proton[18] would have to appear as shown in Fig. 12, according to the Theory of Relativity.

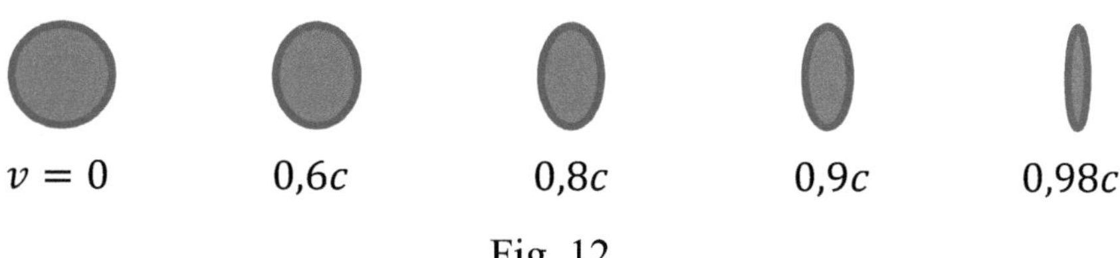

Fig. 12

With the Large Hadron Collider, the most powerful particle accelerator ever realized, protons can be accelerated up to an energy of 14 TeV.

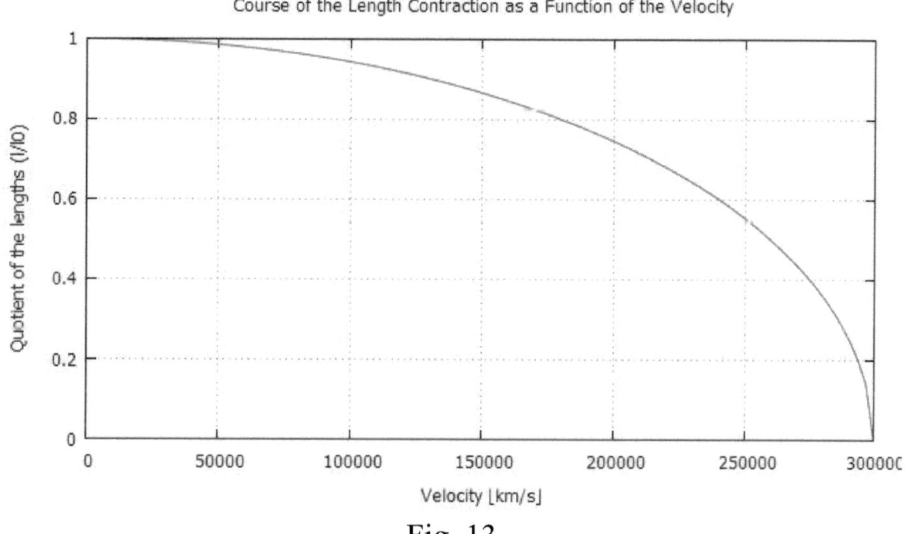

Fig. 13

This energy corresponds to a velocity v_p equal to 99.9999991% of the speed of light.

---

[18] Here, the proton is considered as a classical particle, without considering possible quantum mechanical aspects

At this velocity v_p the proton would have virtually no dimension in the direction of movement.

We will see in Chapter 12 how relativistic length contraction can be derived from a thought experiment using the law of conservation of energy.

In the thirteenth chapter we will then see how, using the relativistic length contraction, we can derive relations for the transformation of spatial and temporal coordinates that turn out to be identical to the Lorentz relations obtained from the postulate of the constancy of the speed of light.

In this section it was shown that, assuming the invariance of simultaneity, the measurements of time intervals taken by observers in relative motion to one another do not agree. Although at this stage of the study we do not yet have the physical means to quantitatively confirm the results of the Theory of Special Relativity, it has meanwhile been shown that time depends on the choice of the frame of reference.

11 The Velocity Addition Formula

Conventional methods of deriving the addition formula for velocities are based on transformations. However, in the fifth Chapter it was stated that in the further course of this work we must forego the use of any transformation, because on the one hand, the Galilean transformation can only be used for low speeds and on the other hand, a transformation for any speed has not yet been derived.

Therefore, we can only rely on the conservation laws of mass, energy and momentum.

To add velocities, we differentiate between two cases: Case 1 for low speeds, and Case 2 for any speeds. Case 1 should clarify the method used.

We assume that, due to a central collision between two particles P_1 of mass m_1 and P_2 of mass m_2, a researcher O observes the formation of a new particle P of mass m_0, which remains at rest with him, as shown in Fig. 14. A second observer O_1 is at rest relative to particle P_1.

Case 1. Addition for low speeds

In order to derive the velocity addition theorem with classical mechanics without using Galilei's transformation, we refer to the thought experiment shown in Fig. 14, using only the laws of conservation of mass and momentum.

From the point of view of the observer O results, for the conservation of mass and momentum before and after the collision:

$$m_0 = m_1 + m_2 \ (a) \quad and \quad m_1 v_1 - m_2 v_2 = 0 \ (b)$$

If v_{12} is the speed of P_2 from the point of view of observer O_1, then the following applies for him:

$$m_2 v_{12} = m_0 v_1 \qquad\qquad (c)$$

Using relation (a) in (c) results in:

$$m_2 v_{12} = m_1 v_1 + m_2 v_1 \qquad (d)$$

From relation (b) we get: $m_1 v_1 = m_2 v_2$ and that inserted into equation (d) results in:

$$v_{12} = v_1 + v_2$$

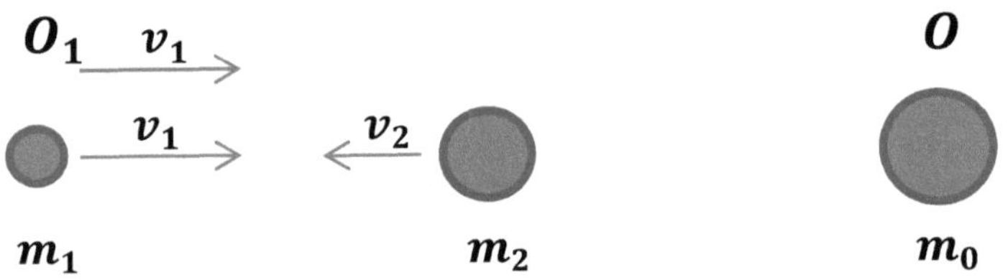

Fig. 14

We will now show how, using a similar method, the relativistic addition formula can be derived for any velocity without using the Lorentz transformation, as required by the traditional interpretation of the Theory of Relativity.

Case 2. Addition for any speed

In order to derive the relativistic addition formula for velocities, the energy and momentum conservation laws are applied to the collision of two dissimilar particles in the next thought experiment.

We will see that the somewhat more complicated calculation method, than that used in Chapter 8 for equal masses and velocities, leads to the relation (9.7).

Assuming that m_{01}, m_{02} are the invariant masses and that v_1, v_2 are the velocities of particles P_1 and P_2 respectively, then the observer O will find the following relationship for the mass m_0 of particle P due to the conservation of energy:

$$m_0 c^2 = m_1 c^2 + m_2 c^2 \qquad \Rightarrow$$

Or using the relation (5.4):

$$m_0 = \frac{m_{01}}{\sqrt{1 - \dfrac{v_1^2}{c^2}}} + \frac{m_{02}}{\sqrt{1 - \dfrac{v_2^2}{c^2}}} \qquad (11.1)$$

On the other hand, the momentum of particle P is zero, because of momentum conservation before and after the collision:

$$\frac{m_{01} v_1}{\sqrt{1 - \dfrac{v_1^2}{c^2}}} - \frac{m_{02} v_2}{\sqrt{1 - \dfrac{v_2^2}{c^2}}} = 0 \qquad (11.2)$$

If v_x/c is replaced by β_x then:

$$\frac{m_{01}}{\sqrt{1 - \beta_1^2}} = \frac{m_{02}}{\sqrt{1 - \beta_2^2}} \frac{\beta_2}{\beta_1} \qquad (11.3)$$

After the right-hand side of equation (11.3) has been inserted into equation (11.1) and the term $\dfrac{m_{02}}{\sqrt{1-\beta_2^2}}$ has been excluded, the following results:

$$m_0 = \frac{m_{02}}{\sqrt{1 - \beta_2^2}} \left(1 + \frac{\beta_2}{\beta_1} \right) \qquad (11.4)$$

Unlike (11.1), Equation (11.4) expresses the value of the invariant mass of the formed particle P as a function of only <u>one</u> of the masses of the colliding particles. We will use this interim result in the course of this derivation.

We now assume a second observer O_1, who is at rest relative to the particle P_1. This can calculate the relative velocity v_{12} between P_1 and P_2 by applying momentum conservation before and after collision of the particles as follows:

Since the observer O_1 is at rest at P_1, the momentum p_1 measured by him before the collision is only that of the particle P_2 with mass m_{02}, which moves towards O_1 with the velocity v_{12}:

$$p_1 = \frac{m_{02} v_{12}}{\sqrt{1 - \dfrac{v_{12}^2}{c^2}}}$$

After the collision, observer O_1 continues to move against particle P at the speed v_1. Therefore, the momentum p_2 measured by O_1 is only that of the particle P with mass m_0:

$$p_2 = \frac{m_0 v_1}{\sqrt{1 - \dfrac{v_1^2}{c^2}}}$$

Because of the law of conservation of momentum, the total momentum before and after the collision must be the same ($p_1 = p_2$). It follows:

$$\frac{m_{02} v_{12}}{\sqrt{1 - \dfrac{v_{12}^2}{c^2}}} = \frac{m_0 v_1}{\sqrt{1 - \dfrac{v_1^2}{c^2}}} \qquad\Rightarrow$$

In order to be able to carry out the following calculations more easily, $^{v_x}/_c$ is now replaced by β_x :

$$\frac{m_{02}\beta_{12}}{\sqrt{1-\beta_{12}^2}} = \frac{m_0\beta_1}{\sqrt{1-\beta_1^2}} \qquad (11.5)$$

(From this point on, see also a derivation in Appendix A IV that is based on the law of conservation of energy).

The term for m_0 from (11.4) is now inserted into equation (11.5):

$$\frac{m_{02}\,\beta_{12}}{\sqrt{1-\beta_{12}^2}} = \frac{m_{02}}{\sqrt{1-\beta_1^2}\sqrt{1-\beta_2^2}}\beta_1\left(1+\frac{\beta_2}{\beta_1}\right) \qquad \Rightarrow$$

Now m_{02} can be shortened on both sides and the bracket with the factor β_1 can be simplified:

$$\frac{\beta_{12}}{\sqrt{1-\beta_{12}^2}} = \frac{\beta_1+\beta_2}{\sqrt{1-\beta_1^2}\sqrt{1-\beta_2^2}} \qquad \Rightarrow$$

To solve the roots, both terms are squared left and right:

$$\frac{\beta_{12}^2}{1-\beta_{12}^2} = \frac{\beta_1^2 + 2\beta_1\beta_2 + \beta_2^2}{1-\beta_1^2 - \beta_2^2 + \beta_1^2\beta_2^2} \qquad \Rightarrow$$

$$\beta_{12}^2 - \beta_{12}^2\beta_1^2 - \beta_{12}^2\beta_2^2 + \beta_{12}^2\beta_1^2\beta_2^2 =$$

$$= \beta_1^2 + 2\beta_1\beta_2 + \beta_2^2 - \beta_{12}^2\beta_1^2 - 2\beta_{12}^2\beta_1\beta_2 - \beta_{12}^2\beta_2^2$$

Finally, the terms $\beta_{12}^2\beta_1^2$ and $\beta_{12}^2\beta_2^2$, which occur with the same sign on both sides of the equation, are eliminated. In addition, the term $2\beta_{12}^2\beta_1\beta_2$ is transferred from the right to the left side of the equation.

It follows:

$$\beta_{12}^2 + 2\beta_{12}^2\beta_1\beta_2 + \beta_{12}^2\beta_1^2\beta_2^2 = \beta_1^2 + 2\beta_1\beta_2 + \beta_2^2 \quad \Rightarrow$$

$$\beta_{12}^2(1 + 2\beta_1\beta_2 + \beta_1^2\beta_2^2) = \beta_1^2 + 2\beta_1\beta_2 + \beta_2^2$$

It is easy to see that the left and right sides of the equation contain squares of binomial formulas.

$$\beta_{12}^2(1 + \beta_1\beta_2)^2 = (\beta_1 + \beta_2)^2 \quad \Rightarrow$$

It follows:

$$\beta_{12} = \frac{\beta_1 + \beta_2}{1 + \beta_1\beta_2}$$

Finally, replacing β_x by v_x/c yields the relativistic velocity addition formula:

$$v_{12} = \frac{v_1 + v_2}{1 + \dfrac{v_1 v_2}{c^2}} \qquad (11.6)$$

Relation (11.6) agrees with Einstein's addition formula for velocities.

Of course, a derivation based on the law of energy conservation leads to the same result (see Appendix A IV).

It should be emphasized once again that the equation (11.6) was derived only from conservation laws. The postulate of the constancy of the speed of light was not used in its derivation.

> Applying the law of energy and momentum conservation to the central collision of two particles makes it possible in the general case to derive the relativistic addition formula of velocities without using the postulate of the constancy of the speed of light Without using the Lorentz transformation.

12 Derivation of length contraction and time dilation

In this chapter we will see how the length contraction can be derived as a function of velocity using the law of conservation of energy.

For this purpose, let us imagine the central collision between two electrons which have a distance $2l$ to each other at time $t = 0$ and are moving toward each other with equal velocities v, as illustrated on the left in Fig. 15.

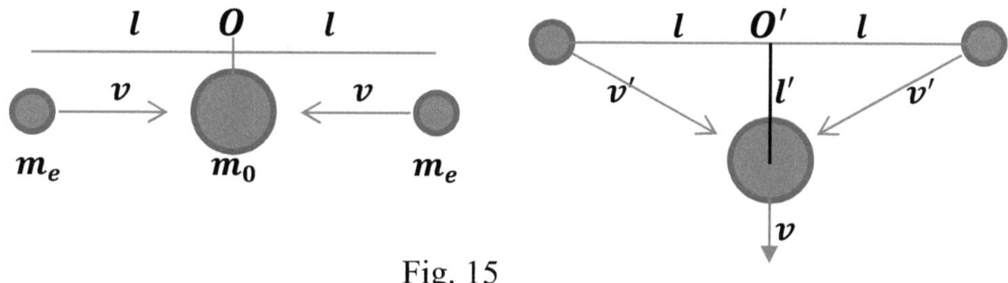

Fig. 15

Let us assume that, as a result of the collision, a new particle forms with the mass m_0.

From the point of view of an observer O, this particle is at the origin of a frame of reference that is at rest relative to him.

Since for observer O the length of the distance traveled by a particle to the collision is equal to l, depending on the time t until the collision, $l = vt$ is valid.

Let us now look at the same thought experiment from the point of view of a second observer O', who rests in a frame of reference that moves at the same velocity v as the particles, but vertically upwards (see the right side of Fig. 15).

Since this motion in y-direction is orthogonal to the direction of motion of the colliding electrons, the observers O and O' measure in x-direction the same length l, the same velocity v and thus, until the collision, also the same time t.[19]

During this time, the observer moves forward in the coordinate system O' by the distance l', i.e., the collision takes place for him on the y-axis shifted backwards by l' (see Figure 15 right).

We assume that, at time $t = 0$, the two origins of the coordinate systems O and O' coincide and that for the observer O' the electrons collide on the Y-axis at the point T at the coordinate l', as shown in Figure 16.

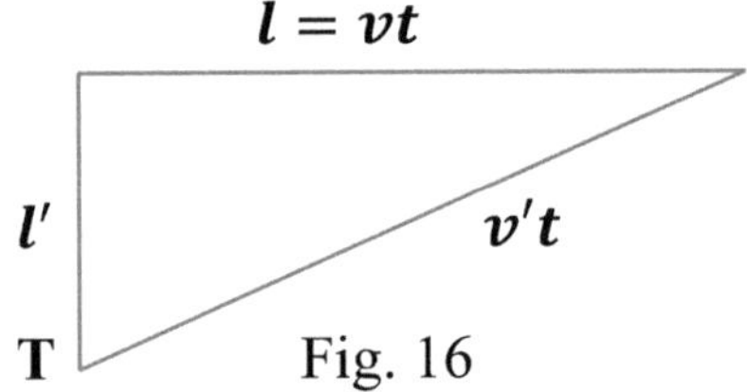

Fig. 16

From Fig. 16 results[20]:

$$v'^2 = v^2 + \frac{l'^2}{t^2} \qquad (12.1)$$

---

[19] In the most general case, an observer needs transformations to describe space and time from the perspective of another observer. These transformations are still unknown in the derivation of the length contraction. Nevertheless, it is certain that for a relative speed $v = 0$ the observers measure the same length and time. This is a two-dimensional case. So that we can orientate ourselves better, I call X the axis on which the electrons move and Y the axis on which observer O' moves. In two dimensions, each of the two observers needs a transformation for the X-axis and a transformation for the Y-axis in order to calculate the lengths from the perspective of the other observer. Both transformations are only dependent on the component v_x and v_y of the relative speed v of the observers. Here only the component of the movement of the particles in the X-direction is needed for both observers to calculate l, v and t. The transformation for the X-axis is sufficient for this. The component v_x of the relative speed v between the observers in the X direction is, however, zero. This is why the transformation for the X-axis provides the same values for lengths and times for both observers.

[20] The time interval t measured by the observers on the X-axis until the collision is necessarily the same as that measured by observer O' on the hypotenuse of the right triangle in Fig. 16. If it were not so, instead of one, observer O' would experience two collisions at different times, the first on the inclined path and the second as a mapping of the first on the X-axis.

In equation (12.1) there are two unknown variables: the length l' and the velocity v'.

Therefore, to calculate the length l', we need a second relation that expresses the velocity v' in the oblique direction as a function of the velocity v only.

We will calculate l' at first for a speed v much lower than the speed of light.

Calculation of l' for $v \ll c$

As Figure 17 shows, the velocity v' of the electrons, measured by observer O' for $v \ll c$, results from the sum of two mutually orthogonal vector components of the same magnitude.

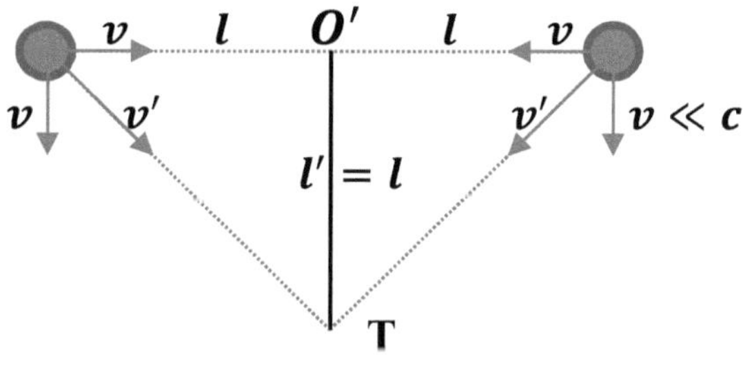

Fig. 17

Thus, the vector v' has an inclination of 45°. Therefore, for $v \ll c$, the electrons meet, from the point of view of O', at the point **T** at

$$l' = l$$

Under these conditions, the vector v' has the value $\sqrt{2}v$, which, however, would assume an impermissible value greater than c for approximately $v > 70\% \, c$.

Thus, the boundary of this classical view becomes clear.

Calculation of l′ for any Velocity:

To calculate the distance l' for any velocities, we need an additional equation for the velocity v'.

It helps that, when calculating v', the conservation of the momentum and the energy must be considered.

This requirement makes it possible to calculate the length l' from a relativistic point of view as follows:

From O point of view, the following applies due to the law of conservation of energy (see the left side of Figure 15):

$$m_0 c^2 = \frac{2m_{0e} c^2}{\sqrt{1 - \dfrac{v^2}{c^2}}} \qquad (12.2)$$

From observer's O' point of view, the formed particle moves further down along the Y-axis after the collision with the velocity v (see the right side of Fig. 15). For O' the result is as follows (see also chapter 18):

$$\frac{m_0 c^2}{\sqrt{1 - \dfrac{v^2}{c^2}}} = \frac{2m_{0e} c^2}{\sqrt{1 - \dfrac{v'^2}{c^2}}} \qquad (12.3)$$

By using (12.2) in (12.3) we get:

$$\frac{2m_{0e} c^2}{1 - \dfrac{v^2}{c^2}} = \frac{2m_{0e} c^2}{\sqrt{1 - \dfrac{v'^2}{c^2}}} \quad ==>$$

$$1 - \frac{v^2}{c^2} = \sqrt{1 - \frac{v'^2}{c^2}} \quad ==>$$

$$1 - 2\frac{v^2}{c^2} + \frac{v^4}{c^4} = 1 - \frac{v'^2}{c^2} \quad ==>$$

$$v'^2 = 2v^2 - \frac{v^4}{c^2} \qquad (12.4)$$

Equation (12.4) is the searched relationship that expresses v' in dependence of v. Because of equation (12.1), now follows:

$$\frac{l'^2}{t^2} = v^2\left(1 - \frac{v^2}{c^2}\right) \qquad (12.5)$$

Since $vt = l$, (12.5) can be rewritten as follows:

$$l'^2 = l^2\left(1 - \frac{v^2}{c^2}\right)$$

And it follows:

$$l' = l\sqrt{1 - \frac{v^2}{c^2}} \qquad (12.6)$$

Relation (12.6) expresses, in accordance with the traditional interpretation of the theory of relativity, the velocity-dependent length contraction in the direction of motion.

In the following chapter we will see how from the relation (12.6) can be derived in a simple way the transformations for spatial and temporal coordinates that are identical to the Lorentz transformations.

From the Lorentz transformations, as we shall see, the relation of time dilation as a function of velocity can be derived.

It takes the following form:

$$t' = \frac{t}{\sqrt{1 - \frac{v^2}{c^2}}} \qquad (12.7)$$

The relation (12.7) expresses, in accordance with the traditional interpretation of the theory of relativity, time dilation as a function of velocity in the direction of motion.

The application of the law of conservation of energy to the collision of two electrons from the point of view of observers in relative motion to each other makes it possible to derive the relation of relativistic length contraction depending on velocity.

13 Transformation of Space and Time Coordinate

In the previous chapter we derived the relativistic length contraction formula as a function of speed using the principle of conservation of energy.

In this chapter we will see how from the relation of the length contraction, the Lorentz transformations for space and time can be derived without having to assume constancy of the speed of light.

Before performing the demonstration, however, we want to briefly describe the traditional method of deriving the transformations.

Demonstration of Lorentz Transformations by the Traditional Method

In 1905 Albert Einstein derived the transformations based on the following assumptions:

1. **Constancy of the speed of light**: The speed of light is the same in all inertial frames of reference, regardless of the motion of the source or observer.

2. **Principle of relativity**: The laws of mechanics are equally valid in all frames of reference in uniform motion.

3. **Linear transformation**: The coordinates between two inertial frames of reference S and S', moving with relative velocity v, are connected by linear equations.

This approach eliminated the need for ether, interpreting the transformations as fundamental properties of space-time.

Mathematical Derivation

Initial assumptions:

- Coordinate systems: S (system at rest) and S' (system moving with velocity v in the direction of x).

- Event: A point in space-time with coordinates (x, t) in S and (x', t') in S'.

Transformation equations:

For linearity and symmetry of the systems we have:

$$x' = \gamma(x - vt) \qquad and \qquad x = \gamma(x' + vt')$$

Calculation of the Lorentz factor

For the calculation of the Lorentz factor, the constancy of the speed of light is applied:

For a flash of light, for which in S we have $x = ct$, the analogous relation $x' = ct'$ must also hold in S'.

Plugging these relations into the transformation equations yields the Lorentz factor:

$$\gamma = \frac{1}{\sqrt{1 - \dfrac{v^2}{c^2}}}$$

This Lorentz factor dominates the relativistic effects at high speeds.

By equating and solving the equations for t', we obtain the transformation for time:

$$t' = \gamma\left(t - \frac{vx}{c^2}\right)$$

Einstein showed that Maxwell's equations are invariant with respect to these transformations, thus resolving contradictions with classical mechanics.

Now let us see how from the relation of the contraction of lengths, Lorentz transformations for space and time can be derived, without having to assume constancy of the speed of light.

We consider two one-dimensional frames of reference in motion relative to each other at constant speed. Two observers O and O' each rest on the coordinate origins of the two frames of reference and measure the time t and t', respectively. The coordinate origins of the two frames of reference coincide at time $t = 0$ for O and $t' = 0$ for O'.

We can accept, as discussed in Chapter 10, that for later times the time measurements t and t' may be different, i.e., we do not assume a priori that $t = t'$, not even for $v \ll c$.

To clearly illustrate the relationships between the frames of reference, we will follow the following rules for the next diagrams:

- Each chart has an observer at rest. A second observer moves along the X axis at speed v.

- The transformation is viewed from the point of view of the resting observer and thus his time is used. The observer at rest is highlighted in **gray** in the diagrams.

- The frame of reference of the moving observer is shown with **dashed lines**.

- The transformation always concerns the calculation of the space coordinate of the moving observer as a function of the space and time coordinates of the observer at rest.

If the relative velocity v between the frames of reference is considerably lower than the speed of light, the observers do not perceive any length contraction in the direction of movement. In Figure 18, the observer O is at rest. t is the time measured by O.

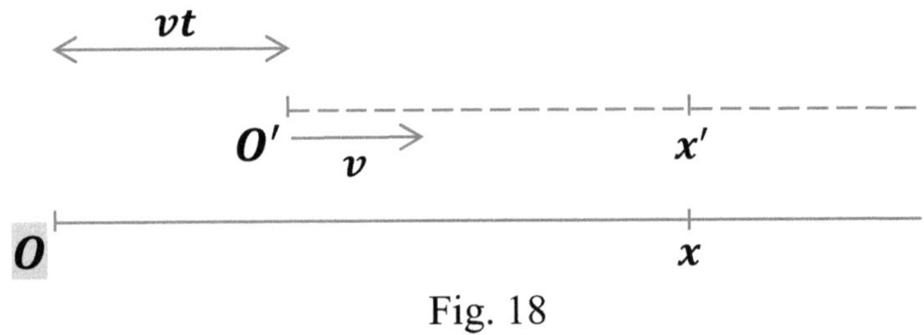

Fig. 18

According to Figure 18, from the point of view of O, the relationships are:

$$x = x' + vt \quad \rightarrow \quad x' = x - vt \quad (13.1)$$

According to Figure 19 from the perspective of observer O' follows:

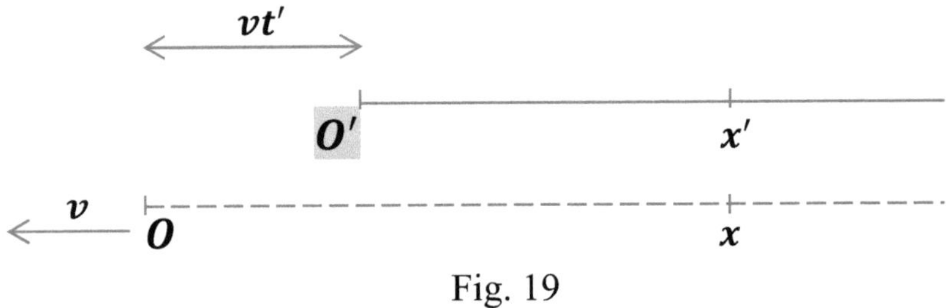

Fig. 19

The corresponding relations are:
$$x' = x - vt' \quad \rightarrow \quad x = x' + vt' \quad (13.2)$$

And if one replaces x' by relation (13.1) then follows:

$$x = x - vt + vt' \quad \rightarrow \quad t = t' \quad (13.3)$$

Relation (13.3) shows for $v \ll c$: The time coordinate is invariant.

The relations (13.1) and (13.3) correspond to the Galilean transformation and are valid in the context of classical mechanics for $v \ll c$:

$$x' = x - vt \quad (13.1); \qquad t = t' \quad (13.3)$$

Now consider the situation from the viewpoint of observer O, this time for any velocity v:

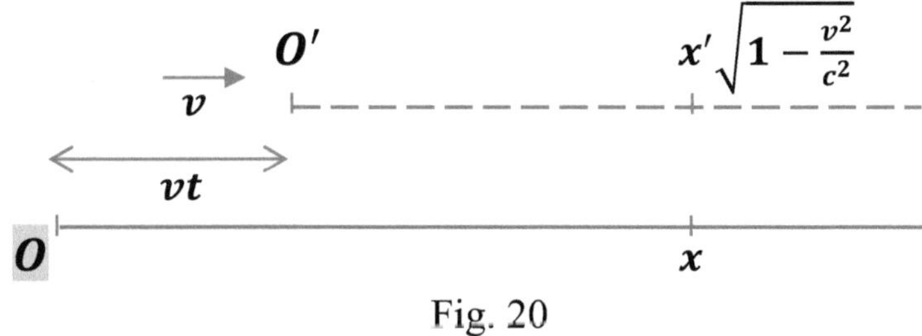

Fig. 20

As proved in Chapter 11, observer O perceives a length contraction of the distance between O' and x', so that for him in the frame of reference O' the point corresponding to x is located at the coordinate $x'\sqrt{1 - \frac{v^2}{c^2}}$, as shown in Figure 20.

From Figure 20 it can be seen that for observers O is valid:

$$x = x'\sqrt{1 - \frac{v^2}{c^2}} + vt \qquad ==>$$

$$x' = \frac{x - vt}{\sqrt{1 - \frac{v^2}{c^2}}} \qquad (13.4)$$

t in (13.4) is the time from the point of view of $\boldsymbol{O}$.

Relation (13.4) represents the value of the space coordinate $\boldsymbol{x'}$ in the frame of reference of the observer $\boldsymbol{O'}$, as a function of the space and time coordinate $(\boldsymbol{x}, \boldsymbol{t})$ of the frame of reference of $\boldsymbol{O}$.

Now let us look at the situation from the viewpoint of observer $\boldsymbol{O'}$:

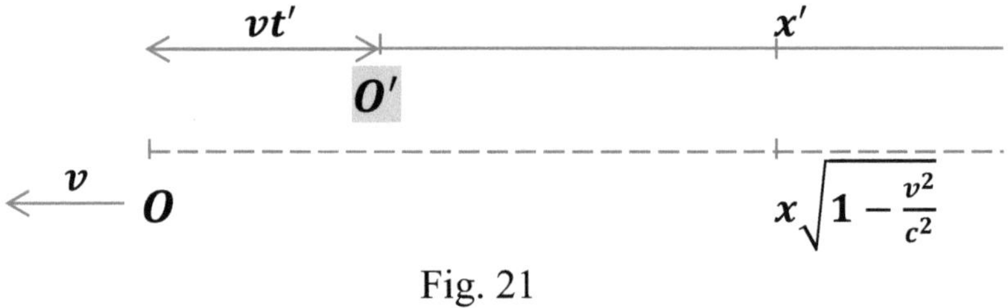

Fig. 21

Observer $\boldsymbol{O'}$ determines a length contraction of the distance between $\boldsymbol{O}$ and $\boldsymbol{x}$, so that for him the corresponding point of $\boldsymbol{x'}$ on the frame of reference $\boldsymbol{O}$ is found at the coordinate $\boldsymbol{x}\sqrt{1-\dfrac{v^2}{c^2}}$, as shown in Figure 21.

From Figure 21 it can also be deduced that for observers $\boldsymbol{O'}$ is valid:

$$\boldsymbol{x'} + \boldsymbol{vt'} = \boldsymbol{x}\sqrt{1-\frac{v^2}{c^2}} \qquad ==>$$

$$\boldsymbol{x} = \frac{\boldsymbol{x'} + \boldsymbol{vt'}}{\sqrt{1-\dfrac{v^2}{c^2}}} \qquad (13.5)$$

It should be noted that $\boldsymbol{t'}$ represents the time from the point of view of $\boldsymbol{O'}$. This is generally different from the time $\boldsymbol{t}$ of Observer $\boldsymbol{O}$, as discussed in Chapter 10.

The relations (13.4) and (13.5) represent the spatial coordinate transformations for any velocities, from the perspective of two observers in relative motion to each other.

Transformation of the Temporal Coordinate:

The transformation for the temporal coordinate can be derived from the relations (13.4) and (13.5).

From relation (13.5) we get:

$$x' = x\sqrt{1 - \frac{v^2}{c^2}} - vt' \qquad (13.6)$$

Relation (13.4) equated with relation (13.6) yields:

$$\frac{x - vt}{\sqrt{1 - \frac{v^2}{c^2}}} = x\sqrt{1 - \frac{v^2}{c^2}} - vt' \quad ==>$$

$$x - vt = x\left(1 - \frac{v^2}{c^2}\right) - vt'\sqrt{1 - \frac{v^2}{c^2}} \quad ==>$$

$$-vt = -x\frac{v^2}{c^2} - vt'\sqrt{1 - \frac{v^2}{c^2}} \quad ==>$$

$$t' = \frac{t - \frac{xv}{c^2}}{\sqrt{1 - \frac{v^2}{c^2}}} \qquad (13.7)$$

Using an analogous method, the following relation can be derived for t as a function of t' and x' :

$$t = \frac{t' + \dfrac{x'v}{c^2}}{\sqrt{1 - \dfrac{v^2}{c^2}}} \qquad (13.8)$$

The coordinate transformations according to (13.4) and (13.5) as well as (13.7) and (13.8) appear in the same form as components of the Lorentz Transformation, which describes the transition between inertial frames of reference for any velocities in the framework of the Theory of Special Relativity.

To conclude, let us compare the two derivation methods described.

With the traditional method, the Lorentz transformations are derived directly from the postulate of the constancy of the speed of light. As a presupposition of Restricted Theory of Relativity, they represent the fundamental equations from which all other relativistic principles are derived. The physicist must reject his innate conception of the absoluteness of space and time before even deriving the laws of mechanics.

With the alternative method, Lorentz transformations are derived at the end of a long chain of demonstrations. They are not the assumption but the result of physical considerations, which are based on the principle of equivalence between mass and energy derived from classical physics. The physicist is confronted with the hypothesis of non-absoluteness of space and time only after deriving the main relativistic laws.

> In this section it was shown how the space and time transformation for any relative velocities of observers can alternatively be derived from the phenomenon of relativistic length contraction.

14 Constancy of the Speed of Light

The speed of light appears as the constant c in many physical formulas. One such formula is the relation of the momentum of the light radiation $p = E/c$. It turns out that with a <u>stationary light source</u> the speed of light remains constant for every frequency from radio waves to gamma rays. This constancy for all frequencies does not violate the laws of classical physics.

This Chapter is not about this concept of constancy, but about the fact that the speed of light remains constant even with a <u>moving light source</u> and between all inertial frames of reference, as the experiment by Michelson and Morley shows.

Conflict with Classical Mechanics

Since this constancy applies to every relative speed between light transmitter and receiver, it is no longer in accordance with the Galilei transformation and thus conflicts with a fundamental principle of classical mechanics. That is also the reason why the concept "constancy of the speed of light in the relativistic sense" eludes the intuition of the human mind all too easily.

But what is so special about this concept to cause a revolution in physics with the known consequences: time dilation, length contraction and the famous paradoxes? In this Chapter we want to try to provide clarity using a concrete example:

Imagine a spaceship that moves relative to an observer O_a (index "a" stands for "outside the spaceship") with a constant, uniform speed v. In the middle of the spaceship, a device simultaneously shoots two bullets k_1 and k_2 at the same speed v_k ($v_k > v$; $v_k \ll c$). Bullet k_1 is shot in the direction of movement of the spaceship, the other bullet k_2 in the opposite direction. From the point of view of an observer O_i (index "i" stands for "inside the spaceship"), who is resting inside the spaceship, the bullets reach the front and back ends of the spaceship simultaneously. But what does the situation look like from the point of view of the observer O_a in relative movement to the spaceship?

During the flight of k_1 the spaceship moved forward. Thus, its front end has moved away from the starting position of k_1. Therefore, from the point of view of observer O_a, which bullet reaches the target first?

The correct answer is that for observer O_a too both bullets reach their targets at the same time, because O_a measures not only a longer path for bullet k_1, but also a higher speed than for bullet k_2.

Because of the relative speed of the space shuttle, observer O_a measures the speeds $v_k + v$ for bullet k_1 and $v_k - v$ for bullet k_2, in accordance with the addition theorem of the speeds from the Galilean transformation. Velocities and lengths then compensate each other and thus the running times of the bullets are the same. So, at low speeds there is simultaneity of the events for all observers.

But let us see what would happen if the device in the middle of the spaceship, instead of the bullets, would emit two light quanta (or photons) in opposite directions. Now let us also assume a much higher velocity for the spaceship than in the previous example.

Analogously for example with the bullets, observer O_a would have to measure the velocities $c + v$, and $c - v$ for the photons, too. But this is not the case: As it is confirmed by experiments, observer O_a, like O_i, measures exactly the same value $c \cong 300000 \ Km/s$ for the velocities of the photons in both directions. This makes the failure of the Galilean transformation obvious and the necessity of a new transformation for space and time recognizable: a transformation is required which takes into account the invariance of the speed of light.

Now the meaning of the term "constancy of the speed of light in the relativistic sense" should be clear: unlike the speed of mass bodies, the speed of light is the same from the point of view of all observers, regardless of their relative movement to the light source and to each other. In our case this leads to the following consequence: for observer O_i, both photons reach the walls of the space shuttle at the same time. For observer O_a, the photon emitted in the direction of travel reaches the destination later because, at the same speed c, it must cover a longer distance than the other photon. So, there is no longer simultaneity of events for both observers. In cases like this one speaks of a relativity of simultaneity. In doing so, the physicist realizes that the concept of time flowing in the same way for all observers, as Newton had imagined, no longer makes sense.

At the end of the nineteenth century, physicists were still intimately familiar with the concept of absolute time and space.

When Michelson and Morley announced the results of their experiments in 1887, the scientists around the world were therefore confronted with a great surprise: The experimental observations conflicted with the principles of mechanics, because the experiment on Michelson's interferometer proved that the speed of light in vacuum is always constant, regardless of the resting or moving state of the light source.

From this knowledge the need arose to lend the natural phenomenon of the constancy of the speed of light the attribute of a fundamental physical postulate.

On the other hand, physicists felt that this postulate was not in accordance with Newton's laws.

As a consequence of this conviction, they refrained from attempting to explain the constancy of the speed of light on the basis of Newton's mechanics within the framework of the laws of classical physics.

Rather, scientists came to believe that it was necessary to develop a new physical theory.

The birth of the Theory of Relativity is therefore closely linked to the assumed incompatibility of Newtonian mechanics with the natural phenomenon of the constancy of the speed of light.

However, we are now able to prove this natural phenomenon theoretically by means of the derivation of the velocity addition carried out in the tenth Chapter.

For a better overview, here is a short summary of the entire path leading to the derivation of the velocity addition formula:

- The relation of the velocity addition (11.6) was derived from the application of conservation laws to the collision of two particles.

- For the energy balance, the total energies of the particles (i.e., the sum of their kinetic and internal energy) were used.

- The formula for the total energy of a particle (7.5) was derived in the seventh Chapter by apply the relation (5.4), which expresses the dependence of mass on velocity.

- In the fifth Chapter, however, it has been shown that the relation of the dependence of mass on velocity (5.4) is a direct consequence of the second law of motion and the Mass–Energy Equivalence Principle.

- The energy-mass equivalence principle was finally derived in Chapters 3 and 4 using only classical physics.

The conclusion of this argumentation is that in this work the proof of the relativistic addition formula for velocities was provided without using the postulate of the constancy of the speed of light.

At this point, the velocity addition formula (relation 11.6) enables us to prove the principle of the constancy of the speed of light in a purely theoretical way.

For this purpose, we assume a light source that moves relative to an observer. He can use equation (11.6) …

$$v_{12} = \frac{v_1 + v_2}{1 + \dfrac{v_1 v_2}{c^2}} \tag{11.6}$$

… to calculate the relative velocity v_l of the emitted light:

If v_1 is replaced by the speed c of the light from a stationary light source and v_2 by the velocity v_q of the light source, then:

$$v_l = \frac{c + v_q}{1 + \dfrac{c v_q}{c^2}} \qquad \Rightarrow$$

$$v_l = \frac{c + v_q}{\dfrac{c + v_q}{c}} \tag{14.1}$$

It is now easy to show that equation (14.1) always assumes the solution $v_l = c$ for any velocity v_q of the light source.

This proves that the speed of light is always the same in each inertial frame of reference, regardless of its state of motion.

Looking more closely at the overall approach that led to this evidence, it can be stated that:

The constancy of the speed of light can be proven purely theoretically, i.e., also without the use of experiments, but for the confirmation of these.

By using the addition formula of velocities, the principle of the constancy of the speed of light can be theoretically proven. From this point of view, the theorem of the constancy of the speed of light is not a postulate, but a principle provable by the laws of physics.

15 The Lorentz Transformation and its application

With the coordinate transformations of Lorentz, which we derived in Chapter 13 by resorting to the principle of conservation of energy and without the postulate of the constancy of the speed of light, we can now solve some important physical applications at high speeds and discuss their results.

We consider an observer O, which is at the origin of a one-dimensional frame of reference characterized by the coordinate x. A second observer O' moves along the X-axis with constant velocity v.

Assuming that for the time $t = 0$ the positions of the observers O and O' match, the following transformations named after Lorentz apply:

$$x' = \frac{x - vt}{\sqrt{1 - \dfrac{v^2}{c^2}}} \quad (15.1) \quad and \quad t' = \frac{t - \dfrac{xv}{c^2}}{\sqrt{1 - \dfrac{v^2}{c^2}}} \quad (15.2)$$

In (15.1) and (15.2), x' and t' give the measurements for the position and the time of a point P from the point of view of the observer O'. They are expressed as a function of the values of the position x and the time t measured by the observer O for the same point.

Solving (15.1) and (15.2) for x and t yields to:

$$x = \frac{x' + vt'}{\sqrt{1 - \dfrac{v^2}{c^2}}} \quad (15.3) \quad and \quad t = \frac{t' + \dfrac{x'v}{c^2}}{\sqrt{1 - \dfrac{v^2}{c^2}}} \quad (15.4)$$

Similarly, in (15.3) and (15.4), x and t give the measurements for the position and time of a point P as seen by the observer O. They are determined by the position x' and the time t' expressed by the observer O' for the same point.

It is important to note that the point **P** does not have to rest in one of the two frames of reference. Consequently, both x and x' can be time dependent.

Since only uniform motions are considered in this discussion, we consider only the following time dependencies of x and x', with constant velocities v_p e v_p':

$$x = x_0 + v_p t \quad (15.5) \qquad and \quad x' = x_0' + v_p' t' \quad (15.6)$$

With the parameters:

v_p is the velocity of **P** in **O**.

x_0 is the position at time point $t = 0$ of **P** in **O**.

v_p' is the velocity of **P** in **O'**.

x_0' is the position at time point $t' = 0$ of **P** in **O'**.

From (15.1) and (15.3) it follows that if $x_0 = 0$ then $x_0' = 0$ and vice versa.

Also, if $v_p' = 0$, then $v_p = v$ and if $v_p = 0$, then $v_p' = -v$ and vice versa.

In summary:

$$x_0 = 0 \iff x_0' = 0 \quad (15.7)$$

$$v_p' = 0 \implies v_p = v \quad (15.8)$$

$$v_p = 0 \implies v_p' = -v \quad (15.9)$$

The following are application examples of the Lorentz Transformation

1) <u>Time of a point at rest in frame of reference O - Time is not absolute</u>

A point **P** rests in the frame of reference O ($v_p = 0$) at the coordinates $x_0 = 0$ and $t = 0$.

We get from relation (15.2):

$$t' = \frac{-\dfrac{x_0 v}{c^2}}{\sqrt{1 - \dfrac{v^2}{c^2}}}$$

For $x_0 > 0$ and $v > 0$ applies: $t' < 0$.

So, while observer O measures time $t = 0$ for point x_0, observer O' sees for the same point a time already in the past.

2) <u>Length of a bar at rest in frame of reference O'- Length contraction</u>

A bar is at rest in the frame of reference of the observer O' and has the length l'. The bar ends have the coordinates $x_1 = 0$ and $x_2 = l$ at the time of the observer O at $t = 0$.

From relation (15.1) results for the bar ends in frame of reference O' :

$$x_1' = 0 \;;\; x_2' = \frac{l}{\sqrt{1 - \dfrac{v^2}{c^2}}} \quad ==>$$

$$x_2' - x_1' = l' = \frac{l}{\sqrt{1 - \dfrac{v^2}{c^2}}} \quad ==>$$

For the bar length in frame O the following applies:

$$l = l' \sqrt{1 - \frac{v^2}{c^2}}$$

Observer O perceives a length contraction for the bar resting in the frame of reference of the observer $\underline{O'}$.

3) <u>Proper time t' of the observer O'- Time dilation</u>

For the observer O' applies in frame O: $x_0 = 0$ and $v_p = v$. From (15.5) follows: $x = vt$. Inserted in (15.2) results in:

$$t' = \frac{t - \dfrac{tv^2}{c^2}}{\sqrt{1 - \dfrac{v^2}{c^2}}} \quad ==> \quad t' = t\sqrt{1 - \frac{v^2}{c^2}} \quad ==> \quad t = \frac{t'}{\sqrt{1 - \dfrac{v^2}{c^2}}}$$

O perceives a time dilation with respect to O'. In other words, the time intervals that pass in frame of reference O' appear extended in frame of reference O.

4) <u>Photon starts in the frame of reference O' at $t' = 0$ and at $x'_0 = 0$ - Constancy of the speed of light</u>

With $x'_0 = 0$ and $v'_p = c$ follows from (15.6): $x' = ct'$. This inserted in (15.3) and (15.4) leads to:

$$x = \frac{ct' + vt'}{\sqrt{1 - \dfrac{v^2}{c^2}}} \qquad and \qquad t = \frac{t' + \dfrac{ct'v}{c^2}}{\sqrt{1 - \dfrac{v^2}{c^2}}}$$

As for $x'_0 = 0$ also $x_0 = 0$, it follows from (15.5) for the velocity of the photon in the frame of reference O:

$$v_p = \frac{x}{t} \quad ==> \quad v_p = \frac{t'(c+v)}{t'\left(1+\frac{v}{c}\right)} \quad ==> \quad v_p = c$$

Despite the relative velocity v between the frames of reference, both observer O and O' will measure from their perspective that the photon moves at the same velocity c.

5) <u>Moving point in the frame of reference O' - Speed addition</u>

A point has the position $x' = 0$ for $t' = 0$ in the frame of reference O'. From (15.6) follows: $x'_0 = 0$ and consequently $x_0 = 0$. The relations (15.5) and (15.6) are reduced to:

$$x = v_p t \qquad and \qquad x' = v'_p t'$$

These are now used in (15.3) and (15.4):

$$v_p t = \frac{v'_p t' + v t'}{\sqrt{1-\frac{v^2}{c^2}}} \quad (a) \qquad and \qquad t = \frac{t' + \frac{v'_p v t'}{c^2}}{\sqrt{1-\frac{v^2}{c^2}}} \quad (b)$$

From (b) results:

$$t' = \frac{t\sqrt{1-\frac{v^2}{c^2}}}{1+\frac{v'_p v}{c^2}}$$

And inserted in (a) we get:

$$v_p = \frac{v'_p + v}{1+\frac{v'_p v}{c^2}} \qquad (c)$$

Equation (c) stands for the relativistic speed addition.

6) <u>Two photons reach observer O' at the same time. Is this also the case from the point of view of O?- Simultaneity</u>

Two photons start in the frame of reference O' at the time $t' = 0$ from the positions $-x_0'$ and x_0' in the direction O'. From (15.6) applies to the first and the second photon:

$$x_1' = -x_0' + ct' \quad und \quad x_2' = x_0' - ct'$$

The two photons simultaneously reach the position of the observer O' at $x' = 0$ after the time $t' = {x_0'}/{c}$.

The starting time of the photons can be calculated from observer O view by relation (15.4).

At the starting point for the first photon applies: $t' = 0$ and $x' = -x_0'$.

From (15.4) follows:

$$t_{01} = \frac{\dfrac{-x_0' v}{c^2}}{\sqrt{1 - \dfrac{v^2}{c^2}}}$$

At the starting point for the second photon applies: $t' = 0$ and $x' = x_0'$.

From (15.4) follows:

$$t_{02} = \frac{\dfrac{x_0' v}{c^2}}{\sqrt{1 - \dfrac{v^2}{c^2}}}$$

One sees: for observers O, the photons do not start simultaneously. The first photon starts earlier.

At what point in time do the photons reach the position of the observer O from the point of view of the observer O'?

We have seen that the following applies to the first and the second photon:

$$x_1' = -x_0' + ct' \quad and \quad x_2' = x_0' - ct'$$

Inserted in (15.4), we get for the times of the two photons:

$$t_1 = \frac{t' + \dfrac{(-x_0' + ct')v}{c^2}}{\sqrt{1 - \dfrac{v^2}{c^2}}} \quad and \quad t_2 = \frac{t' + \dfrac{(x_0' - ct')v}{c^2}}{\sqrt{1 - \dfrac{v^2}{c^2}}}$$

Since it is the time $t' = {x_0'}/{c}$ that passes for both photons in frame O', we get:

$$t_1 = t_2 = \frac{x_0'}{c\sqrt{1 - \dfrac{v^2}{c^2}}} = \frac{x_0'}{\sqrt{c^2 - v^2}}$$

The photons thus reach the position of observer O' simultaneously for observer O as well.

For observer O the observer O' moves at velocity v. Therefore, the first photon must travel a longer distance up to the observer O' than the second at the same speed c. Nevertheless, the photons reach the observer O' at the same time, because from the observer's O point of view, the first photon starts earlier than the second.

.

16 Derivation of the relativistic Doppler Effect

The relativistic Doppler Effect plays a key role in the traditional derivation of the Theory of Relativity.

In fact, after deriving it from the laws of electrodynamics, Einstein used this principle in his fourth paper in 1905 to support the hypothesis of the dependence of the inertia of bodies on their energy content.

In this chapter we will see how a simple demonstration of the relativistic Doppler Effect can be made by applying the principles of conservation of energy and momentum to the physical process of pair annihilation.

For this, we again refer to phases II and III of the physical process considered in the fourth Chapter, where the annihilation of one particle and the subsequent emission of two photons are described.

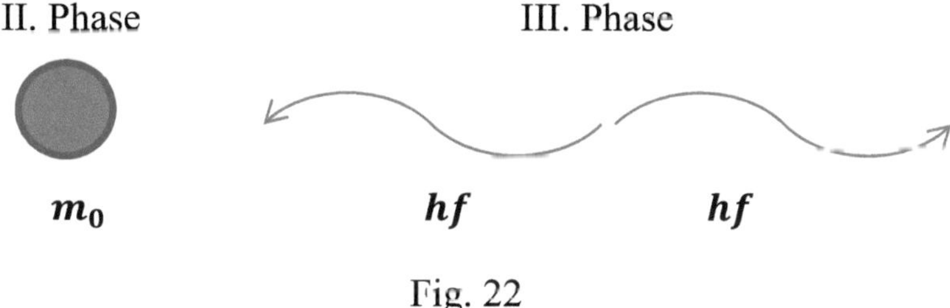

Fig. 22

Consider an observer moving in the same direction of a photon at a low speed relative to the particle.

As already stated in the third Chapter, he will then measure the following frequency shift due to the optical Doppler Effect [21]:

$$f' = f\left(1 \pm \frac{v}{c}\right) \qquad (16.1)$$

---

[21] The following relation (16.1) is experimentally verifiable with the measuring instruments available to physicists nowadays.

However, if the velocity of the observer is close to that of the light, then it turns out that the expression (16.1) is no longer correct.

Frequency Variation in the most General Case

Therefore, to calculate the frequency change as a function of the velocity in the general case, we will apply the law of conservation of energy and momentum to phases II and III of the described natural phenomenon.

Since the entire mass of the particle transforms into the energy of the photons, the following applies:

$$m_0c^2 = 2hf \qquad\qquad (16.2)$$

Where f is the frequency assigned to the photon, h is Planck's constant and c is the speed of light in a vacuum.

If f_1 and f_2 represent the frequencies measured by the observer in or against the direction of motion, the following equation can be established, because of the energy conservation, before and after the annihilation of the particle:

$$mc^2 = \frac{m_0c^2}{\sqrt{1 - \dfrac{v^2}{c^2}}} = hf_1 + hf_2 \qquad\qquad (16.3)$$

The application of the law of conservation of momentum also results in:

$$mv = \frac{m_0v}{\sqrt{1 - \dfrac{v^2}{c^2}}} = \frac{hf_1}{c} - \frac{hf_2}{c} \qquad\qquad (16.4)$$

If the expression $2hf/c^2$ from equation (16.2) is substituted into formulas (16.3) and (16.4) instead of m_0, the following system of equations results from algebraic simplification:

$$\begin{cases} f_1 + f_2 = \dfrac{2f}{\sqrt{1 - \dfrac{v^2}{c^2}}} \\[4ex] f_1 - f_2 = \dfrac{2f\dfrac{v}{c}}{\sqrt{1 - \dfrac{v^2}{c^2}}} \end{cases}$$

Resolution of the System of Equations

Solving for f_1 and f_2, we get:

$$f_1 = f\,\frac{(1 + \frac{v}{c})}{\sqrt{1 - \frac{v^2}{c^2}}} \quad ; \quad f_2 = f\,\frac{(1 - \frac{v}{c})}{\sqrt{1 - \frac{v^2}{c^2}}}$$

Or in a compact form:

$$f' = f\left(1 \pm \frac{v}{c}\right)\gamma \qquad\qquad (16.5)$$

Where γ represents the so-called Lorentz factor.

It should be noted that the relations (16.1) and (16.5) are identical except for this factor, and that the classical formula (16.1) represents a limiting case of the relativistic relation (16.5) for $v \ll c$ (and consequently for $\gamma = 1$).

By simple algebraic transformations, the following expressions finally result for the frequency shift in, or contrary to the direction of movement, dependent on the velocity:

$$f_1 = f\sqrt{\frac{c + v}{c - v}} \quad ; \quad f_2 = f\sqrt{\frac{c - v}{c + v}} \qquad (16.6)$$

The equations (16.6) agree with those derived for the optical Doppler Effect by the traditional demonstration method of the Theory of Special Relativity.

In Figures 23 and 24, the violet curve shows the classical course and the green curve the relativistic course of the frequency shift.

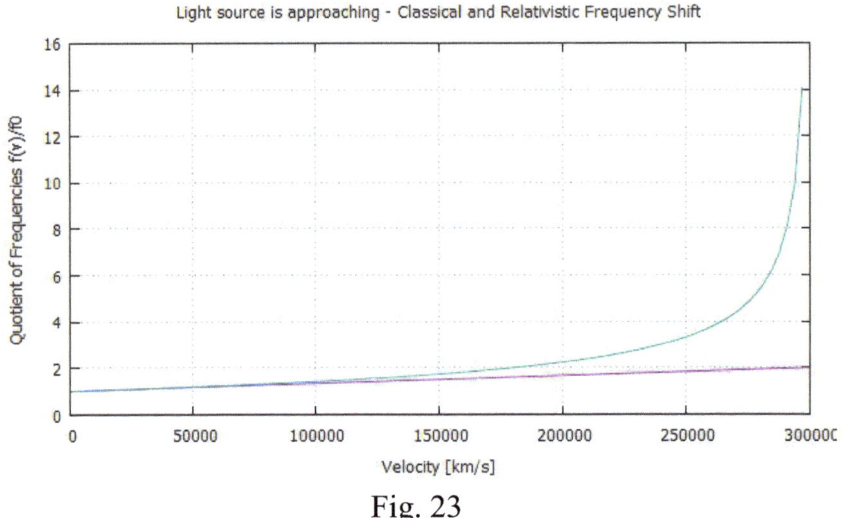

Fig. 23

According to the classical formula of the frequency shift (16.1) for a light source approaching the observer (Fig. 23), the frequency can maximally double up to the speed of light, while according to the relativistic formula (16.6), for high speeds of the light source, the frequency tends towards infinity.

However, if the light source moves away from the observer, the difference between the curves of formulas (16.1) and (16.6) is no longer relevant (Fig. 24).

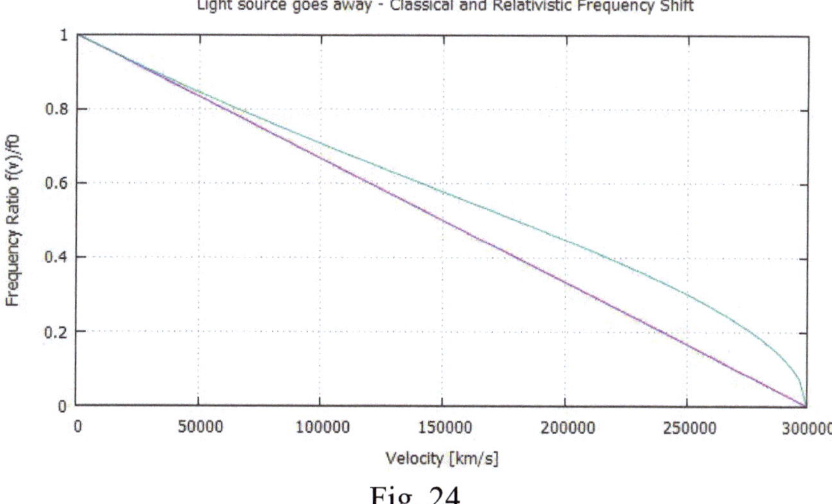

Fig. 24

In this case, it is also interesting to note that not only the relativistic relation (16.6), but also the classical (16.1) indicates that the speed of light cannot be exceeded. This is because the equation $f' = f(1 - v/c)$ with $v > c$ yields a negative and therefore impermissible frequency value.

In this respect, however, the classical equation (16.1) does not possess the sharpness of the relativistic relation (16.6), which is defined mathematically neither for $v > c$ nor for $v = c$.

The application of the momentum and energy conservation law to the experimental observation of electron positron annihilation makes it possible to determine the frequency shift of the electromagnetic radiation as a function of the velocity of the emitting source.

119

17   Dependence of Acceleration on Velocity

It has already been pointed out in Chapter 2 that Newton's second law of motion in connection with the relativistic mass formula (5.4) leads to the relation of relativistic acceleration (see Appendix A II).

To calculate the acceleration, the second law of motion is also used in this section, which, as stated in the first Chapter, can be expressed in the more general case by the following relation:

$$\vec{F} = \frac{d(m\vec{v})}{dt} \quad \Leftrightarrow \vec{F} = \vec{v}\frac{dm}{dt} + m\frac{d\vec{v}}{dt} \qquad (17.1)$$

To derive the speed dependence of the acceleration, two proofs are presented below:

In the first proof, for the sake of simplicity, the case is considered in which a body moves in the same direction as the force acting on it.

This first proof is based on a purely scalar calculation and is suitable for describing the movement of particles in linear accelerators.

The second proof uses a vector calculation to derive the two vector components (longitudinal and transverse) of acceleration.

This second proof covers the situations in which a body does not move in the same direction as the force acting on it, as e.g. occurs by the movement of celestial bodies.

Both cases are suitable to show that the second law of motion represents the fundamental relation for calculating the relativistic acceleration.

Case 1. Scalar calculation.

In this derivation we will limit ourselves to rectilinear motions in which, as is the case in linear accelerators, the path of the particles proceeds in the same direction as the force.

In this case, relation (17.1) can be used in scalar form:

$$F = v\frac{dm}{dt} + m\frac{dv}{dt} \qquad (17.2)$$

Since the path is in the same direction of force, relation (5.1) (see Chapter 5) can be used for infinitesimal work:

$$Fds = c^2 dm \qquad (5.1)$$

From (5.1) follows:

$$dm = \frac{Fvdt}{c^2} \qquad (17.3)$$

After substituting (17.3) and the relativistic relation $m = m_0/\sqrt{1 - v^2/c^2}$ into (17.2) and reducing variables we obtain:

$$F = \frac{v^2}{c^2}F + \frac{m_0}{\sqrt{1 - \dfrac{v^2}{c^2}}}\frac{dv}{dt}$$

It follows:

$$F = \frac{m_0 a}{\left(1 - \dfrac{v^2}{c^2}\right)^{\frac{3}{2}}} \qquad (17.4)$$

From equation (17.4) the formula of linear acceleration as a function of velocity can be derived:

$$a = \frac{F}{m_0}\left(1 - \frac{v^2}{c^2}\right)^{\frac{3}{2}} \qquad\qquad (17.5)$$

It is easy to see that (17.5) can be simplified to scalar form of relation (1.1) if $v \ll c$.

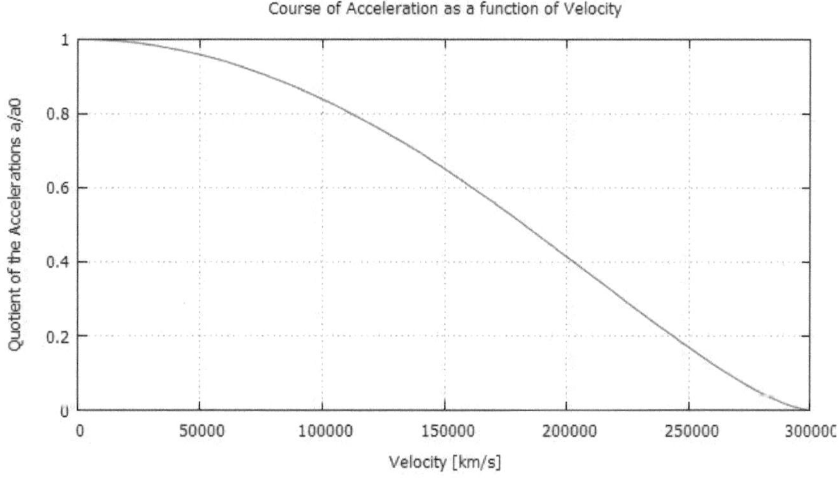

Fig. 25

Equation (17.5) shows that with constant force and increasing speed, the acceleration gradually decreases and tends towards zero near the speed of light (Fig. 25).

This result agrees with the experimental observations that can be made in the particle accelerators.

It is interesting to note that the relation (17.5) can also be derived from the differential equation (7.3) used for the calculation of kinetic energy in Chapter 7:

$$Fds = \frac{m_0 v}{\left(1 - \dfrac{v^2}{c^2}\right)^{\frac{3}{2}}} dv \qquad\qquad (7.3)$$

The relation (17.5) is obtained with the following steps:

(i)   In (7.3) replace the infinitesimal distance ds by the product of the velocity with the time differential $\boldsymbol{vdt}$

(ii)   replace the differential of the speed $\boldsymbol{dv}$ with the product of the acceleration with the time differential $\boldsymbol{adt}$,

(iii)   shortening once and

(iv)   finally, solve the equation for acceleration $\boldsymbol{a}$.

Case 2. Vector calculation.

If the path does not run in the same direction of the force, the relation (5.1) for the infinitesimal work must be written as follows:

$$\vec{F} \cdot \vec{ds} = \vec{F} \cdot \vec{v}dt = c^2 dm \qquad\qquad (17.6)$$

Where $\vec{F} \cdot \vec{ds}$ represents the scalar product of the force with the infinitesimal displacement vector.

From (17.6) it follows:

$$\frac{dm}{dt} = \frac{\vec{F} \cdot \vec{v}}{c^2} \qquad\qquad (17.7)$$

Relation (17.7) and the relativistic mass formula $m = m_0 / \sqrt{1 - v^2/c^2}$ are now inserted in (17.1):

$$\vec{F} = \frac{\vec{F} \cdot \vec{v}}{c^2}\vec{v} + \frac{m_0}{\sqrt{1 - \frac{v^2}{c^2}}}\frac{d\vec{v}}{dt} \qquad (17.8)$$

In the following vector calculation, we will use the parallel component (longitudinal component) and the perpendicular component (transverse component) to the velocity $\vec{v}$ for all vectors:

$$\vec{F} = \begin{pmatrix} F_L \\ F_T \end{pmatrix}; \qquad \frac{d\vec{v}}{dt} = \vec{a} = \begin{pmatrix} a_L \\ a_T \end{pmatrix}; \qquad \vec{v} = \begin{pmatrix} v \\ 0 \end{pmatrix}$$

Equation (17.8) leads to:

$$\begin{pmatrix} F_L \\ F_T \end{pmatrix} = \frac{1}{c^2}\left[\begin{pmatrix} F_L \\ F_T \end{pmatrix} \cdot \begin{pmatrix} v \\ 0 \end{pmatrix}\right]\begin{pmatrix} v \\ 0 \end{pmatrix} + \frac{m_0}{\sqrt{1 - \frac{v^2}{c^2}}}\begin{pmatrix} a_L \\ a_T \end{pmatrix} \qquad \Rightarrow$$

$$\begin{pmatrix} F_L \\ F_T \end{pmatrix} - \frac{F_L v}{c^2}\begin{pmatrix} v \\ 0 \end{pmatrix} = \frac{m_0}{\sqrt{1 - \frac{v^2}{c^2}}}\begin{pmatrix} a_L \\ a_T \end{pmatrix} \qquad \Rightarrow$$

Since F_L has the same direction as v it follows:

$$\begin{pmatrix} F_L \\ F_T \end{pmatrix} - \frac{v^2}{c^2}\begin{pmatrix} F_L \\ 0 \end{pmatrix} = \frac{m_0}{\sqrt{1 - \frac{v^2}{c^2}}}\begin{pmatrix} a_L \\ a_T \end{pmatrix} \qquad \Rightarrow$$

$$\begin{pmatrix} (1 - \frac{v^2}{c^2})F_L \\ F_T \end{pmatrix} = \frac{m_0}{\sqrt{1 - \frac{v^2}{c^2}}}\begin{pmatrix} a_L \\ a_T \end{pmatrix} \qquad \Rightarrow$$

From this it follows for the longitudinal and the transverse component of the acceleration:

$$a_L = \frac{F_L}{m_0}\left(1 - \frac{v^2}{c^2}\right)^{\frac{3}{2}} \quad ; \quad a_T = \frac{F_T}{m_0}\left(1 - \frac{v^2}{c^2}\right)^{\frac{1}{2}} \qquad (17.9)$$

Figure 26 shows the curves of the longitudinal and the transverse vector components of the relativistic acceleration as a function of the speed. These relations also agree with those that were derived with the "mathematical formalism of the theory of relativity".

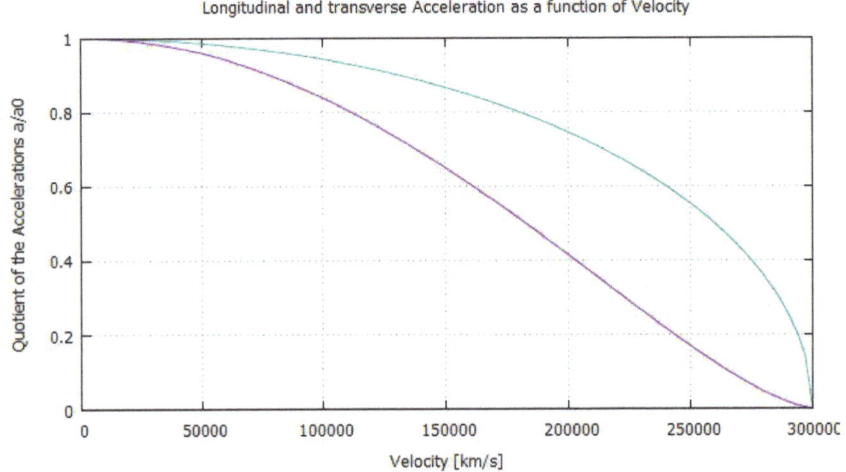

Fig. 26

The definition of the second law of dynamics in its general form makes it possible to derive the expression of acceleration as a function of speed. The curves in Figures show that for velocities close to the speed of light, the acceleration approaches zero, as confirmed by experimental observations.

18 Addition of orthogonal velocities

With the alternative derivation method used in this work, we have so far limited ourselves to proving and confirming well-known relativistic formulas. In this chapter we want to go further, showing that with the same method it is possible to derive relativistic relations not yet known.

Using the principle of conservation of energy we now want to derive the relativistic addition formula for orthogonal velocities, which must be used, for example, to give an answer to problems of the following kind.

Consider the following Astronomical Problem:

A galaxy moves away with a velocity equal to 80% of the speed of light in a direction inclined 45° from the axis joining an observer with the center of the galaxy itself.

A star in the galaxy moves with the same velocity, but in a direction orthogonal to it, so that the resultant of the two velocities has the same direction as the axis of conjunction between observer and galaxy.

We pose the following question:

How fast does the star move away from the observer?

For classical mechanics, in this case, the velocity of the star would be obtained by applying the Pythagorean Theorem to the modulus of the component vectors that are orthogonal to each other.

This value of the velocity would turn out to be greater than that of light, with the consequence that the star would no longer be visible.

What, however, would be the value of the star's velocity relative to the observer according to the Theory of Special Relativity?

To give an answer to this question, we need the relativistic addition formula for orthogonal velocities, which we will derive in this chapter.

Demonstration of the Formula for Two Velocities

Consider the central collision between two electrons moving toward each other at the same velocities v_x as shown on the left in Figure 27.

Suppose that as a result of the collision a new particle of mass m_0 is formed that, from the point of view of an observer O, remains in the origin of a frame of reference at rest with him.

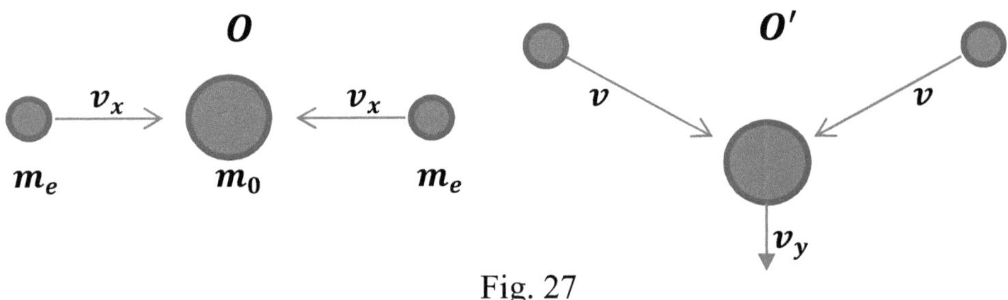

Fig. 27

Let us now consider the same thought experiment from the point of view of a second observer O' moving upward at the velocity v_y orthogonal to the direction of motion of the colliding electrons (see in the figure on the right).

From the point of view of observer O, because of the law of conservation of energy, the following results (see in figure left):

$$m_0 c^2 = \frac{2 m_{0e} c^2}{\sqrt{1 - \frac{v_x^2}{c^2}}} \qquad (18.1)$$

Where m_{0e} is rest mass of the electron.

From the point of view of the observer O', the particle formed after the collision moves down the Y axis with the velocity v_y (see right in the figure) and so for him it results:

$$\frac{m_0 c^2}{\sqrt{1 - \frac{v_y^2}{c^2}}} = \frac{2m_{0e} c^2}{\sqrt{1 - \frac{v^2}{c^2}}} \tag{18.2}$$

Using equation (18.1) in relation (18.2) we obtain:

$$\frac{2m_{0e} c^2}{\sqrt{1 - \frac{v_x^2}{c^2}} \sqrt{1 - \frac{v_y^2}{c^2}}} = \frac{2m_{0e} c^2}{\sqrt{1 - \frac{v^2}{c^2}}} \quad \Rightarrow$$

$$\left(1 - \frac{v_x^2}{c^2}\right)\left(1 - \frac{v_y^2}{c^2}\right) = 1 - \frac{v^2}{c^2} \quad \Rightarrow$$

$$1 - \frac{v_x^2}{c^2} - \frac{v_y^2}{c^2} + \frac{v_x^2 v_y^2}{c^4} = 1 - \frac{v^2}{c^2}$$

Hence it follows:

$$v^2 = v_x^2 + v_y^2 - \frac{v_x^2 v_y^2}{c^2} \tag{18.3}$$

The relation (18.3) expresses the relativistic addition formula for orthogonal velocities.

It is easy to see that the relation (18.3) for $v_x \ll c$ and $v_y \ll c$ reduces to $v^2 = v_x^2 + v_y^2$, just as it is known to be valid for the addition of orthogonal vectors.

If either or both components are equal to c, then it results: $v = c$.

For other arbitrary values of v_x and v_y, according to relation (18.3), v never exceeds the speed of light, as would be expected from the theory of relativity. Of this last statement we provide a proof by contradiction.

Reductio ad Absurdum

Using for the speed of light the absolute value 1 let us assume by proof by contradiction that:

$$v_x^2 + v_y^2 - v_x^2 v_y^2 > 1 \quad \Rightarrow$$

$$v_x^2 + v_y^2 - v_x^2 v_y^2 - 1 > 0 \quad \Rightarrow$$

$$v_x^2 (1 - v_y^2) - (1 - v_y^2) > 0 \quad \Rightarrow$$

$$(1 - v_y^2)(1 - v_x^2) < 0$$

Since v_y^2 and v_x^2 are both less than or at most equal to 1 this last inequality is never true.

Quod erat domonstrandum.

I am not aware that the relation (18.3) is known in the context of the traditional derivation method of the Theory of Relativity.

Let us now use relation (18.3) to solve the astronomical problem we posed at the beginning of the chapter.

Using for the speed of light the absolute value 1 we obtain:

$$v^2 = 0.8^2 + 0.8^2 - 0.8^4$$

From which we get: $v \cong 0.933$.

From relation (18.3) we can derive the formula for the relativistic composition of three orthogonal velocities.

Suppose that v_1, v_2 and v_3 are the moduli of three orthogonal velocities, the resultant velocity of which we want to calculate.

In a first step of the demonstration we calculate the resultant v_{12} of the velocities v_1 and v_2 with the formula (18.3):

$$v_{12}^2 = v_1^2 + v_2^2 - \frac{v_1^2 v_2^2}{c^2} \qquad (18.4)$$

Since the velocity v_3 is itself orthogonal to v_{12}, from the composition of these two velocities with (18.3), we obtain the resultant v_{123} of the three velocities v_1, v_2 and v_3:

$$v_{123}^2 = v_{12}^2 + v_3^2 - \frac{v_{12}^2 v_3^2}{c^2} \qquad (18.5)$$

Substituting v_{12}^2 in (18.5) with the result of (18.4), after simplifying, we obtain:

$$v_{123}^2 = v_1^2 + v_2^2 + v_3^2 - \frac{v_1^2 v_2^2}{c^2} - \frac{v_1^2 v_3^2}{c^2} - \frac{v_2^2 v_3^2}{c^2} + \frac{v_1^2 v_2^2 v_3^2}{c^4} \qquad (18.6)$$

The expression (18.6) represents the relativistic addition for orthogonal velocities in 3D.

It can easily be seen that if one of the velocities is equal to zero, (18.6) takes the same form as (18.3).

If one, two or all velocities are equal to the speed of light c, then it turns out that the resultant v_{123} is also equal to c and this is also the maximum value it can take.

Summary

The alternative derivations of Einstein, Rohrlich and Cester show that the famous formula $E = mc^2$, usually considered a relativistic equation, is a simple consequence of the interaction between electromagnetic radiation and matter.

Thus derived, the Mass–Energy Equivalence Principle provides to the Newtonian mechanics with the missing relation that allows the integration of the differential equation derived from the second law of motion in its more general form:

$$\begin{cases} dE = v^2 dm + mv\,dv \\ dE = c^2 dm \end{cases} \quad \Rightarrow$$

$$c^2 dm = v^2 dm + mv\,dv \qquad (5.2)$$

The integration of (5.2) yields the first important relationship for the inertial mass of a body as a function of speed:

$$m = \frac{m_0}{\sqrt{1 - \dfrac{v^2}{c^2}}} \qquad (5.4)$$

Using this relation and the laws of conservation of energy and momentum, and without using relativistic axiom-based hypotheses, it is possible to derive successively other important relativistic formulas:

- The expression of kinetic energy in the more general case:

$$E_k = \frac{m_0 c^2}{\sqrt{1 - \dfrac{v^2}{c^2}}} - m_0 c^2 \qquad (7.4)$$

- The equation of the total energy of the point mass:

$$mc^2 = \frac{m_0 c^2}{\sqrt{1 - \dfrac{v^2}{c^2}}} = E_k + m_0 c^2 \qquad (7.5)$$

- The relationship between energy, mass and momentum illustrated by the so-called E-p-m triangle:

$$E = mc^2 = c\sqrt{p^2 + m_0^2 c^2} \qquad (8.1)$$

- The velocity addition formula:

$$v_{12} = \frac{v_1 + v_2}{1 + \dfrac{v_1 v_2}{c^2}} \qquad (11.6)$$

- The relativistic length contraction and time dilation depending on the speed:

$$l' = l\sqrt{1 - \frac{v^2}{c^2}} \qquad (12.6) \qquad t' = \frac{t}{\sqrt{1 - \dfrac{v^2}{c^2}}} \qquad (12.7)$$

- The alternative derivation of the Lorentz transformation for space and time:

$$x' = \frac{x - vt}{\sqrt{1 - \dfrac{v^2}{c^2}}} \qquad (13.4) \qquad t' = \frac{t - \dfrac{xv}{c^2}}{\sqrt{1 - \dfrac{v^2}{c^2}}} \qquad (13.7)$$

- The independence of the speed of light v_l from the relative speed v_q between emitting light source and observer:

$$v_l = \frac{c + v_q}{1 + \dfrac{v_q}{c}} = c \qquad (14.1)$$

- The frequency shift for electromagnetic waves at any speed by approaching and receding of the light source:

$$f' = f\sqrt{\frac{c+v}{c-v}} \qquad f' = f\sqrt{\frac{c-v}{c+v}} \qquad (16.6)$$

- The relations of the longitudinal and transverse components of the relativistic acceleration as a function of the speed:

$$a_L = \frac{F_L}{m_0}\left(1 - \frac{v^2}{c^2}\right)^{\frac{3}{2}} \quad and \quad a_T = \frac{F_T}{m_0}\left(1 - \frac{v^2}{c^2}\right)^{\frac{1}{2}} \qquad (17.9)$$

All these formulas agree with those derived from the axioms of the Theory of Special Relativity or rather, with the use of the Lorentz transformations.

And not only that, but… With the alternative derivation method used in this work, it is possible to derive relativistic relations not yet known. One example is the composition of orthogonal velocities.

For two velocities:

$$v^2 = v_x^2 + v_y^2 - \frac{v_x^2 v_y^2}{c^2} \qquad (18.3)$$

And for three velocities:

$$v_{123}^2 = v_1^2 + v_2^2 + v_3^2 - \frac{v_1^2 v_2^2}{c^2} - \frac{v_1^2 v_3^2}{c^2} - \frac{v_2^2 v_3^2}{c^2} + \frac{v_1^2 v_2^2 v_3^2}{c^4} \qquad (18.6)$$

Final Word

The alternative derivations examined in this book show how the results of Relativity Theory can be confirmed on the basis of classical physics.

It is shown that, by deducing several statements of the Theory of Special Relativity, Newtonian mechanics covers a much wider range of applications than is normally assumed.

The Mass-Energy Equivalence Principle -

The relation of the dependence of the mass on the speed -

The equation of the kinetic and total energy of the point mass -

The relativistic triangle that geometrically illustrates the relationship between energy, mass and momentum of a point mass -

The velocity addition formula according with the theorem of Einstein -

The relation of relativistic length contraction and time dilation depending on speed –

The alternative derivation of the Lorentz transformations -

The electromagnetic frequency shift for any velocities -

The relativistic acceleration, dependent on speed ...

In the world of physics, these relations are considered strictly relativistic and can therefore be proven only by the Lorentz transformation.

However, as shown above, they can also be derived starting from classical physics by applying Newton's second law of motion, the mass-energy equivalence principle and the conservation laws of energy and momentum.

Examples

I. Example - Application of conservation laws to electromagnetic absorption

Fig. I shows a body of mass m_0 that absorbs a photon of frequency f at a certain point in time.

Because of the absorption, the body changes from rest to movement with velocity v.

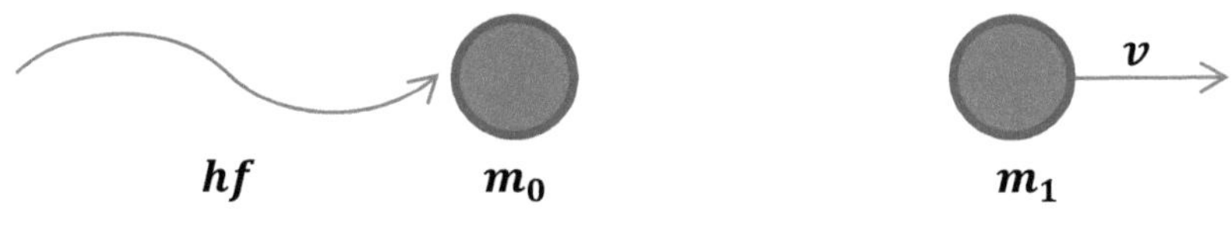

Fig. I

By applying conservation laws before and after the absorption, we find:

For the momentum:

$$\frac{hf}{c} = m_1 v \tag{I.1}$$

For the energy:

$$m_0 c^2 + hf = m_1 c^2 \tag{I.2}$$

From (I.1) we have $m_1 = hf/cv$ and that one used in (I.2) leads to:

$$m_0 c^2 + hf = \frac{hf}{cv} c^2 \tag{I.3}$$

We solve the equation (I.3) for v:

$$v = \frac{chf}{m_0 c^2 + hf}$$

We now assume, absurdly, that the velocity v could be greater than or equal to the speed of light:

$$\frac{chf}{m_0 c^2 + hf} \geq c \qquad\qquad (I.4)$$

From (I.4) follows:

$$m_0 c^2 \leq 0$$

From this the following statement can be made: the hypothesis that the velocity of a body is greater than or equal to the speed of light presupposes the untenable assumption that the mass of the body is equal to or less than zero.

II. Example - Application of conservation laws to electromagnetic emission

Fig. II shows a body of mass m_0 emitting a photon of frequency f at a given time.

Because of the emission, the body changes from the resting to the moving state with the speed v.

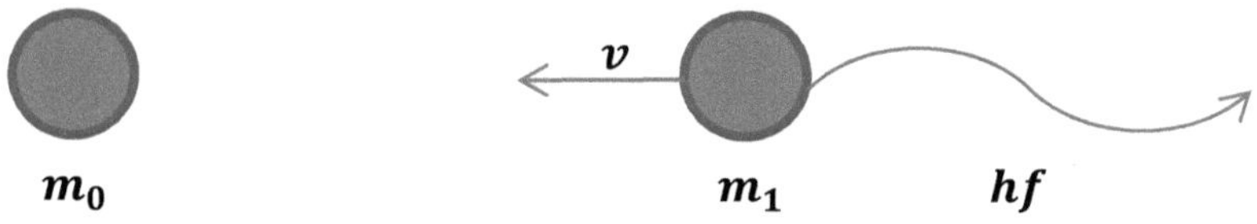

v

m_0 $\qquad\qquad$ m_1 $\qquad$ hf

Fig. II

By applying the conservation laws before and after the emission, we note:

For the momentum:

$$0 = \frac{hf}{c} - m_1 v \qquad\qquad (\text{II. 1})$$

For the energy:

$$m_0 c^2 = m_1 c^2 + hf \qquad\qquad (\text{II. 2})$$

From (II.1) we have $m_1 = hf/cv$ and that one used in (II.2) leads to:

$$m_0 c^2 = \frac{hf}{cv} c^2 + hf \qquad\qquad (\text{II. 3})$$

We solve the equation (II.3) for v:

$$v = \frac{chf}{m_0 c^2 - hf}$$

If we consider that the speed of a body is always lower than the speed of light, we can make the following inequality:

$$\frac{chf}{m_0c^2 - hf} < c$$

It follows:

$$hf < \frac{1}{2}m_0c^2 \qquad\qquad (\text{II}.4)$$

From (II.4) it can be concluded that if a single photon is emitted, its energy hf is always lower than half the internal energy of the emitting body.

Historia Operis

During my years as a student, I started to calculate an alternative derivation of the equation $E = mc^2$.

I knew that this equation is related to the Theory of Relativity. But I also knew that it holds for all phenomena in the context of classical physics unchanged[22]. Why should she then be provable only "relativistically" and not "classically"?

I first tried to prove the formula with classical physics using the law of conservation of energy and the physical properties of electromagnetic radiation. But I failed again and again in the calculation of the derivation.

Since the relation $E = mc^2$ concerns energy, I tried to solve the problem by relying on evidence based on energy balance. And so my attempts were always unsuccessful until the day I read the following sentences in Wikipedia about the famous formula:

"An alternative version of Einstein's thought experiment was proposed by Fritz Rohrlich (1990), who based his reasoning on the Doppler Effect. Like Einstein, he considered a body at rest with mass M. If the body is examined in a frame moving with nonrelativistic velocity v, it is no longer at rest and in the moving frame it has momentum P = Mv."

After reading these lines I suddenly realized: In the derivation I should not use the energy balance and not the law of conservation of energy, but an equation based on the conservation of momentum!

Based on this information, I began to search for the solution myself.

---

22 In all exothermic reactions, a mass decrease (or mass defect) of the reaction products compared to the starting materials can be determined. This mass defect can be explained by a conversion of mass into energy according to the Mass-Energy Equivalence Principle. This applies to both nuclear and chemical reactions. Many of these reactions can only be described by classical physics.

Proof of the Mass-Energy Equivalence $E = mc^2$

Finally, I succeeded to the goal.

Chapters 3 and 4 list three alternative derivations of the equivalence principle E-M. Two of these derivations are based on the optical Doppler effect or the frequency shifts of electromagnetic radiation.

This sentence, unlike the acoustic Doppler Effect, has two important properties that, in a sense, already link it to the Theory of Relativity:

$$f' = f(1 \pm v/c) \qquad (16.1)$$

First, v in the frequency shift formula (16.1) represents the relative velocity between the light source and the observer. That is, it does not matter if the *light source, the observer, or both are moving. Every observer can consider himself as being at rest.*

Thus, the classical relation (16.1) for the propagation of electromagnetic waves, as well as the relativistic formula, does not include any physical dependence on a carrier medium, let alone on an absolute space.

Furthermore, equation (16.1) shows that for $v > c$ the frequency would have a negative and therefore impermissible value.

Proof of the Dependence of Mass on Speed

After this first step, a few years passed until the day I realized that the relation of the principle of equivalence energy-mass can be used to solve the differential equation of mechanical work (see relation 1.5).

This led to the alternative derivation of the dependence of mass on the velocity that is described in the fifth Chapter of this book.

Remarkable in this derivation is the appearance of the Lorentz factor $1/\sqrt{1 - v^2/c^2}$ in the mass formula (see relation 5.4). This factor is always considered to be an exclusive result of the Lorentz transformations, but they are not used in this derivation.

- Just coincidence? - and - Is this just a unique sense of achievement? - I asked myself then.

But one thing was clear to me: With this derivation it had become apparent that there was more potential in the "Lex Secunda" of Newton than I had previously suspected.

In its complete form, that is, with both terms (see relation 2.1), Newton's law had led to this particularly important relativistic relation.

Was that any indication that you could do any more with it?

I wondered if the second term $\vec{v}\dfrac{dm}{dt}$ of relation (2.1) not at all enables the *missing link* between Newtonian and relativistic mechanics.

To verify this, I had to try to apply Newton's law to what I had realized so far.

I now had two important relativistic formulas at my disposal, but they had been derived from a purely classical point of view.

The **first** was the relation of the equivalence of mass and energy:

$$\Delta E = \Delta m c^2 \qquad (2.6)$$

The **second** was the formula of mass dependence on speed:

$$m = m_0/\sqrt{1 - v^2/c^2} \qquad (5.4)$$

I knew that the integration of the differential equation $dE_k = m_0 v\,dv$ from the second law of motion (see relation 1.2) leads to the calculation of the kinetic energy in classical mechanics:

$$E_k = m_0 \int_0^v v\,dv = \frac{1}{2} m_0 v^2 \quad (7.1)$$

Proof of Relativistic Kinetic Energy

It was clear to me that the kinetic energy (7.1) can only apply to low velocities and immutable mass, because it is derived solely from the first term $\boldsymbol{m_0 v\,dv}$ of the differential equation of the work. However, in its full form, it must be defined as follows:

$$dE_k = mv\,dv + v^2 dm \quad (1.5).$$

If I wanted to get a valid relation of kinetic energy for any velocities, I had to integrate (1.5). However, this task is not readily possible because $\boldsymbol{m}$ in (1.5) is speed-dependent unlike m_0 in (1.2).

But now this dependence could be removed by using the two additional formulas (2.6) and (5.4).

Their substitution in (1.5) then led to the differential equation (7.3), whose integration yielded the formula of relativistic kinetic energy.

This was yet another particularly important formula of the Theory of Relativity derived from the "Lex Secunda".

There was no doubt about it:

The application of the second law of motion with its full content led directly to the Theory of Relativity and the second term $\vec{v}\,\dfrac{dm}{dt}$ of (1.5) seemed to be the *link* to it.

Proof of Relativistic Acceleration

A similar procedure as for kinetic energy was used for the derivation of the acceleration for any velocities (see Chapter 17)

The result was the confirmation of another relativistic formula without the use of the Lorentz transformation, which is currently the basis of Relativity Theory.

As an important link in the chain of proof up to the constancy of the speed of light, all that was missing was the derivation of the relativistic addition formula for velocities.

At first, I introduced a thought experiment, in which two electrons traveling at the same speed collide.

Proof of the Relativistic Addition Theorem of Velocities

By using the mass formula (5.4) and the energy conservation law, I then provided a first proof of the addition formula for equal velocities (see Chapter 9).

After this, with a similar method as in Chapter 9, I started looking for an alternative derivation of the relativistic addition formula for velocities, but this time for any velocities.

The calculation led to a rather complicated algebraic derivation with many terms, which gave hardly any hope to reach the correct result.

In the middle of the calculation, however, the equations became increasingly simplified as if by magic (see equations after relation 11.5) and then led to the relativistic formula of the velocity addition.

I finally arrived!

With this last demonstration it had been possible to show that the constancy of the speed of light, as the foundation of the Theory of Special Relativity, is not a postulate but a principle provable by the laws of physics.

Thus, in this way, it had been possible to trace a completely convincing and intuitive path, linking Newtonian mechanics to relativistic mechanics.

And yet I was not quite satisfied.

With the sequence of alternative derivations, I had indeed achieved the principle of the constancy of the speed of light. According to this principle the Lorentz transformations can be derived by two different methods, as Max Born shows.

The first method uses a rather complicated and therefore difficult to follow geometric derivation.

The second method uses a much simpler algebraic method, which is based on the rather arbitrary assumption that the desired transformation is linear and therefore differs from the Galilean transformation only because of a factor to be calculated (it is the Lorentz factor).

Both derivations do not agree with the method used in my work, which instead uses conservation principles of energy and momentum to prove all other relativistic formulas.

The problem was related to the following question:

How can length contraction or time dilation be derived for a frame of reference in relative motion?

Proof of Length Contraction

In Chapter 10 I had shown that there must be a dependence of time on speed, since two observers do not agree on measured time intervals.

However, the thought experiment used did not make it possible to determine this dependence correctly.

Starting from the arbitrary assumption that the observers measure the same lengths, the formulas obtained for time did not agree with the relativistic ones.

On the other hand, I was faced with a seemingly insoluble task, because to calculate the dilation of time correctly, I had to know the contraction of space, which, however, could only be calculated if the temporal dilation was known beforehand.

To solve the problem, I had to use a thought experiment in which the observers could agree on one of the measurements, either of time or length.

This happens, if the relative motion of the observers takes place on an axis orthogonal to the one on which the one-dimensional mechanical processes we want to examine take place, and thus taking the same measurements for distances, velocities and times.

The thought experiment described in Chapter 12 allows observers, moving orthogonally to the collision course of the particles, to define a "common" time interval that elapses until both particles collide.

The consideration of a possible time dilatation is thus omitted and the length contraction in the direction of motion of the observers results directly from the energy and momentum conservation laws.

Using the formula of length contraction as a function of velocity, it is then very easy to derive the Lorentz transformation, as shown in Chapter 13.

This last step thus completes the alternative derivation of the Theory of Special Relativity.

Appendix

An elementary derivation of $E = mc^2$

Fritz Rohrlich

Department of Physics, Syracuse University, Syracuse, New York 13244–1130

(Received 6 March 1989; accepted for publication 12 April 1989)

The equality $E = mc^2$ is derived in a fashion suitable for presentation in an elementary physics course for nonscience majors. It assumes only 19th-century physics and knowledge of the photon.

Einstein's original derivation of the relation between the inertia and the energy content of a body[1] assumes the knowledge of the relativistic Doppler effect. It was done after his seminal paper on special relativity, but 17 years before Compton's experiments of 1922. Had it been done before 1905 and had the particle properties of the photon already been known at that time, the following derivation could conceivably have been carried out.

One starts with the following four simple assumptions.

(1) The Newtonian formulas for the kinetic energy and the linear momentum of a free body of mass m and speed v, $mv^2/2$ and mv. Correspondingly, one assumes $(v/c)^2 \ll 1$ throughout.

(2) The laws of conservation of energy and momentum, but not the law of conservation of mass believed before 1905 to be valid.

(3) The Doppler effect, which has been known since the first half of the 19th century: Radiation (whether it be sound waves or electromagnetic waves) of speed c and frequency v (when the observer is at rest relative to the source) will be perceived to have that frequency altered by a factor $1 + v/c$ $(1 - v/c)$ when the observer moves relative to the source with speed v and in a direction toward it (away from it).

(4) Electromagnetic radiation (in particular visible light) as produced by a source at rest consists of quanta (photons) that have particle properties: Radiation of frequency v consists of photons of energy hv and momentum hv/c. h and c are *constants*.

If these properties of photons had been known to a 19th-century physicist, these statements would have been the accepted truth of the day. Accept them therefore as the basis for an analysis of the following physical process that could be considered as a thought experiment, but which is not beyond realization.

A source of radiation emits two photons simultaneously while remaining at rest in some (Newtonian!) inertial reference frame R. Conservation of momentum requires these two photons to have equal and opposite momenta, and therefore equal frequencies v. Therefore, they also have equal energies hv. Conservation of energy requires that the internal energy of the source diminishes by an amount

$$\Delta E = 2hv. \tag{1}$$

Assume now that this process is viewed from a different reference frame R', which is moving uniformly relative to the rest frame of the source and in such a way that the source is seen to move with speed v in the same direction as one of the photons. Conservation of momentum then requires the momentum of the source before emission p_i' to be equal to the momentum of the source after emission p_f' together with the two momenta of the photons:

$$p_i' = p_f' + \left(\frac{hv}{c}\right)\left(1 + \frac{v}{c}\right) - \left(\frac{hv}{c}\right)\left(1 - \frac{v}{c}\right).$$

The loss in source momentum, $\Delta p'$, is therefore

$$\Delta p' = (2hv/c^2)v. \tag{2}$$

But the source in reference frame R is at rest both before and after emission; in frame R' it must therefore have the same speed v both before and after emission. Now, according to assumption (1) above, the Newtonian formula for momentum is the product of mass times speed. The momentum loss of the source is thus found to *require* a change in mass Δm times v; and that change of mass is found to be

$$\Delta m = 2hv/c^2. \tag{3}$$

Conservation of energy further requires that the initial energy of the source E_i' be equal to its final energy E_f', together with the energies of the two photons:

$$E_i' = E_f' + hv(1 + v/c) + hv(1 - v/c),$$

or

$$\Delta E' = 2hv = \Delta E. \tag{4}$$

Thus the energy loss ΔE of the source is the same in both reference frames, R and R'.

Inserting Eq. (4) into Eq. (3), the change of mass is found to be

$$\Delta m = \Delta E/c^2. \tag{5}$$

One is thus forced to conclude that the emission of the two photons reduces the mass of the source, and this mass loss amounts to an energy loss of $\Delta E = \Delta mc^2$. The equivalence of inertial mass loss and energy loss has thus been derived from the above four assumptions.

One can go one step further and assume that *all* of the mass of the source is used up by emitting photons of large enough frequency; it must be so that $hv = m_i c^2/2$, where m_i is the initial mass of the source. That mass then disappears, and its energy is present in the two photons that have total energy $E = m_i c^2$. Therefore, the mass m_i must have been associated with that amount of energy.

This concludes the elementary derivation. One can add to it several layers of sophistication. The simplest one is to permit the source (as seen from R') to move at an arbitrary angle α relative to one of the photons. This adds a factor $\cos \alpha$ in the Doppler effect. Momentum conservation in R' then requires two equations, one for the parallel and one for the perpendicular components of the momenta. The end result is of course the same.

Another modification would be to assume the relativistic Doppler effect and the relativistic expressions for the linear momentum of the source. One can still maintain $\alpha = 0$ at first. This results in the relativistic relation between the (rest) mass and the total energy (mass energy plus kinetic energy) of the source. If a finite angle α is also assumed, one returns to the assumptions underlying Einstein's original derivation, and our assumption (4) is no longer necessary to derive Eq. (5).

This article was motivated by the criticism[2] of a faulty derivation in my recent book.[3]

The mass formula $m = m_0/\sqrt{1-\frac{v^2}{c^2}}$ is used in the equation $F = d(mv)/dt$ of Newton's second law of motion. The differentiation leads directly to the formula of relativistic acceleration for any velocities. Thus, the assertion is refuted that the second law of motion can only be used with constant mass.

$$F = m_0 \frac{d}{dt}\, v\left(1 - \frac{v^2}{c^2}\right)^{-0.5}$$

$$F = m_0 \frac{dv}{dt}\left(1 - \frac{v^2}{c^2}\right)^{-0.5} - 0.5 m_0 v \left(1 - \frac{v^2}{c^2}\right)^{-1.5}\left(-2\frac{v}{c^2}\right)\frac{dv}{dt}$$

$$F = m_0 \left(1 - \frac{v^2}{c^2}\right)^{-0.5}\frac{dv}{dt} + m_0 \left(1 - \frac{v^2}{c^2}\right)^{-0.5}\left(1 - \frac{v^2}{c^2}\right)^{-1.0}\frac{v^2}{c^2}\frac{dv}{dt}$$

$$F = m_0 \left(1 - \frac{v^2}{c^2}\right)^{-0.5}\left(1 + \left(1 - \frac{v^2}{c^2}\right)^{-1.0}\frac{v^2}{c^2}\right)\frac{dv}{dt}$$

$$F = m_0 \left(1 - \frac{v^2}{c^2}\right)^{-0.5}\left(1 + \frac{\frac{v^2}{c^2}}{1 - \frac{v^2}{c^2}}\right)\frac{dv}{dt}$$

$$F = m_0 \left(1 - \frac{v^2}{c^2}\right)^{-1.5}\frac{dv}{dt}$$

$$a = \frac{F}{m_0}\left(1 - \frac{v^2}{c^2}\right)^{\frac{3}{2}}$$

A III. (Referenced in Chapter 9)

For the thought experiment described in Figure 9, a further derivation is performed here which uses the law of conservation of momentum.

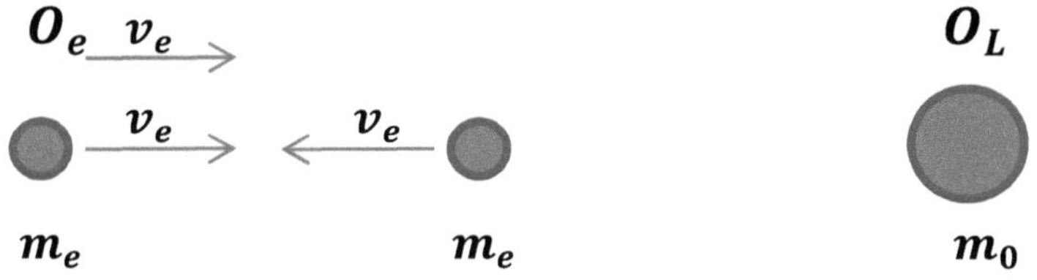

Fig. 9

$$\frac{m_{0e}v_{ee}}{\sqrt{1-\dfrac{v_{ee}^2}{c^2}}} = \frac{2m_{0e}v_e}{1-\dfrac{v_e^2}{c^2}} \qquad \Rightarrow$$

$$\frac{v_{ee}^2}{1-\dfrac{v_{ee}^2}{c^2}} = 4\,\frac{v_e^2}{1-2\dfrac{v_e^2}{c^2}+\dfrac{v_e^4}{c^4}} \qquad \Rightarrow$$

$$v_{ee}^2 - 2\frac{v_{ee}^2 v_e^2}{c^2} + \frac{v_{ee}^2 v_e^4}{c^4} = 4v_e^2 - 4\frac{v_{ee}^2 v_e^2}{c^2} \qquad \Rightarrow$$

$$v_{ee}^2 + 2\frac{v_{ee}^2 v_e^2}{c^2} + \frac{v_{ee}^2 v_e^4}{c^4} = 4v_e^2 \qquad \Rightarrow$$

$$v_{ee} + \frac{v_{ee} v_e^2}{c^2} = 2v_e \qquad \Rightarrow$$

$$v_{ee} = \frac{2v_e}{1+\dfrac{v_e^2}{c^2}} \qquad (9.6)$$

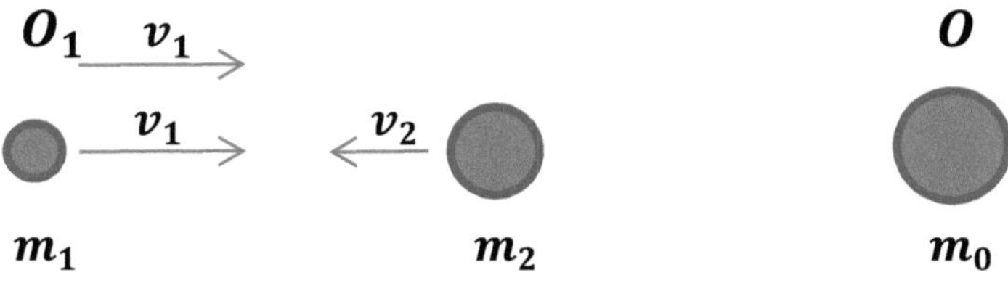

Fig. 14

For the thought experiment described in Figure 14, a further derivation is performed here which uses the law of conservation of energy. From the point of view of observer O_1 applies:

- m_1 is in rest,

- m_2 moves with relative velocity v_{12} which is derived here for any velocities (for low velocities as already known $v_{12} = v_1 + v_2$ is valid),

- m_0 moves at the speed v_1.

That's why we can start with the following equation:

$$m_{01}c^2 + \frac{m_{02}c^2}{\sqrt{1 - \beta_{12}^2}} = \frac{m_0 c^2}{\sqrt{1 - \beta_1^2}} \qquad \text{(AIV.1)}$$

The terms for m_{01} from equation (11.3)

$$\frac{m_{01}}{\sqrt{1 - \beta_1^2}} = \frac{m_{02}}{\sqrt{1 - \beta_2^2}} \frac{\beta_2}{\beta_1} \qquad \text{(11.3)}$$

and for m_0 from equation (11.4)

$$m_0 = \frac{m_{02}}{\sqrt{1 - \beta_2^2}} \left(1 + \frac{\beta_2}{\beta_1}\right) \qquad \text{(11.4)}$$

are inserted into equation (AIV.1):

$$\frac{m_{0z}}{\sqrt{1-\beta_2^2}}\frac{\beta_2}{\beta_1}\sqrt{1-\beta_1^2}+\frac{m_{0z}}{\sqrt{1-\beta_{12}^2}}=\frac{m_{0z}}{\sqrt{1-\beta_2^2}\sqrt{1-\beta_1^2}}\left(1+\frac{\beta_2}{\beta_1}\right)$$

$$\frac{\frac{\beta_2}{\beta_1}\sqrt{1-\beta_1^2}\sqrt{1-\beta_{12}^2}+\sqrt{1-\beta_2^2}}{\sqrt{1-\beta_2^2}\sqrt{1-\beta_{12}^2}}=\frac{\left(1+\frac{\beta_2}{\beta_1}\right)}{\sqrt{1-\beta_2^2}\sqrt{1-\beta_1^2}}$$

$$\frac{\beta_2}{\beta_1}(1-\beta_1^2)\sqrt{1-\beta_{12}^2}+\sqrt{1-\beta_2^2}\sqrt{1-\beta_1^2}=\sqrt{1-\beta_{12}^2}+\frac{\beta_2}{\beta_1}\sqrt{1-\beta_{12}^2}$$

$$\frac{\beta_2}{\beta_1}\sqrt{1-\beta_{12}^2}-\beta_1\beta_2\sqrt{1-\beta_{12}^2}+\sqrt{1-\beta_2^2}\sqrt{1-\beta_1^2}=\sqrt{1-\beta_{12}^2}+\frac{\beta_2}{\beta_1}\sqrt{1-\beta_{12}^2}$$

$$\sqrt{1-\beta_2^2}\sqrt{1-\beta_1^2}=(1+\beta_1\beta_2)\sqrt{1-\beta_{12}^2}$$

$$(1-\beta_2^2)(1-\beta_1^2)=(1+\beta_1\beta_2)^2(1-\beta_{12}^2)$$

$$1-\beta_1^2-\beta_2^2+\beta_1^2\beta_2^2=1+2\beta_1\beta_2+\beta_1^2\beta_2^2-(1+\beta_1\beta_2)^2\beta_{12}^2$$

$$-\beta_1^2-\beta_2^2=2\beta_1\beta_2-(1+\beta_1\beta_2)^2\beta_{12}^2$$

$$(1+\beta_1\beta_2)^2\beta_{12}^2=(\beta_1+\beta_2)^2$$

$$\beta_{12}=\frac{\beta_1+\beta_2}{1+\beta_1\beta_2}$$

$$v_{12}=\frac{v_1+v_2}{1+\frac{v_1v_2}{c^2}}\qquad(11.6)$$

Q.E.D.

A V. (Referenced in Chapter 1)

Physical Quantity	Classical Mechanics	Relativistic Mechanics
Mass	$m = m_0$	$m = \dfrac{m_0}{\sqrt{1 - \dfrac{v^2}{c^2}}}$
Momentum	$p = m_0 v$	$p = \dfrac{m_0 v}{\sqrt{1 - \dfrac{v^2}{c^2}}}$
Longitudinal Acceleration Transverse Acceleration	$a_L = \dfrac{F_L}{m_0}$ $a_T = \dfrac{F_T}{m_0}$	$a_L = \dfrac{F_L}{m_0}\left(1 - \dfrac{v^2}{c^2}\right)^{\frac{3}{2}}$ $a_T = \dfrac{F_T}{m_0}\left(1 - \dfrac{v^2}{c^2}\right)^{\frac{1}{2}}$
Addition of Velocities	$v_{12} = v_1 + v_2$	$v_{12} = \dfrac{v_1 + v_2}{1 + \dfrac{v_1 v_2}{c^2}}$
Kinetic Energy	$E_k = \dfrac{1}{2} m_0 v^2$	$E_k = \dfrac{m_0 c^2}{\sqrt{1 - \dfrac{v^2}{c^2}}} - m_0 c^2$
Frequency Shift Source $\rightarrow \leftarrow$ Observer $\leftarrow$ Source Observer $\rightarrow$	$f' = f\left(1 + \dfrac{v}{c}\right)$ $f' = f\left(1 - \dfrac{v}{c}\right)$	$f' = f\sqrt{\dfrac{c + v}{c - v}}$ $f' = f\sqrt{\dfrac{c - v}{c + v}}$
Length Contraction	$l' = l$	$l' = l\sqrt{1 - \dfrac{v^2}{c^2}}$

Time Dilation	$t' = t$	$t' = \dfrac{t}{\sqrt{1 - \dfrac{v^2}{c^2}}}$
Space Transformation	$x' = x - vt$	$x' = \dfrac{x - vt}{\sqrt{1 - \dfrac{v^2}{c^2}}}$
Time Transformation	$t' = t$	$t' = \dfrac{t - \dfrac{xv}{c^2}}{\sqrt{1 - \dfrac{v^2}{c^2}}}$

The relativistic formulas of **Mass**, **Acceleration** and **Kinetic Energy** can be derived directly from Newton's Second Law of Motion in connection with the Mass-Energy Equivalence Principle $E = mc^2$.

Bibliography of cited works

John Stache – Einsteins Annus mirabilis

Max Born – Die Relativitätstheorie Einsteins

Franz von Krbek – Grundlagen der Mechanik

Wolfgang Pauli – Teoria della Relatività

Albert Einstein – Come io vedo il mondo – La teoria della relatività

Brandes & Czerniawski – Spezielle und Allgemeine Relativitätstheorie